A Guide to Penrose Tilings

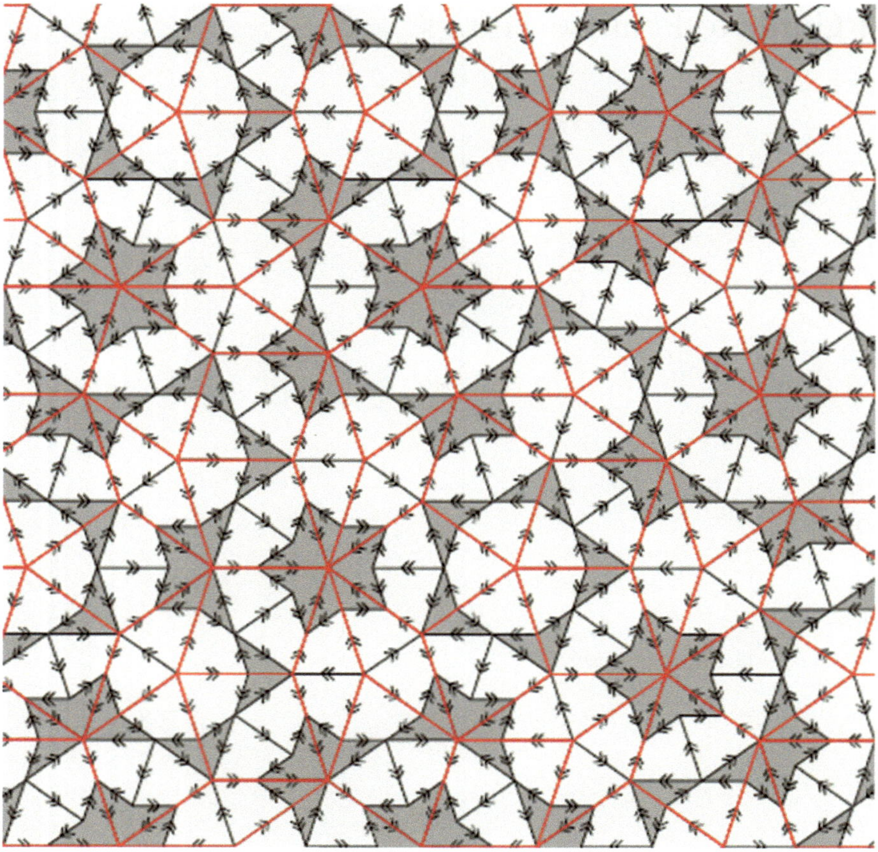

Francesco D'Andrea

A Guide to Penrose Tilings

Springer

Francesco D'Andrea
Department of Mathematics
and Applications "Renato Caccioppoli"
University of Naples Federico II
Napoli, Italy

ISBN 978-3-031-28427-4 ISBN 978-3-031-28428-1 (eBook)
https://doi.org/10.1007/978-3-031-28428-1

This Springer imprint is published by the registered company Springer Nature Switzerland AG
The registered company address is: Gewerbestrasse 11, 6330 Cham, Switzerland

Preface

The aim of this book is to provide an elementary introduction, complete with detailed proofs, to the celebrated tilings of the plane discovered by Sir Roger Penrose in the '70s. The book covers many aspects of Penrose tilings, including the study of the space parameterizing Penrose tilings from the point of view of Connes' Noncommutative Geometry.

I am indebted to my students and colleagues for their comments on a preliminary version of this text. Special thanks go to Prof. Giovanni Landi for his suggestions regarding the last chapter of the book.

All the images in this book were created in *LaTeX* with the TikZ package.

Napoli, Italy Francesco D'Andrea

Contents

1 Introduction 1

2 Tilings and Puzzles 7
- 2.1 Preliminary Notions 7
- 2.2 Square Tilings, Puzzles and Matching Rules 11
- 2.3 Symmetries of Tilings 16
- 2.4 Substitution Tilings 21
- 2.5 The Extension Theorem 23
  - 2.5.1 The Hausdorff Distance 25
  - 2.5.2 Group Actions 27
  - 2.5.3 The Extension Theorem 28

3 Robinson Triangles 33
- 3.1 Decomposition and Existence of Tilings 35
  - 3.1.1 The Cartwheel 38
  - 3.1.2 The Sun and the Star 39
  - 3.1.3 The Golden Triangle and the Golden Gnomon 41
- 3.2 Vertex Neighborhoods 43
- 3.3 Composition and Aperiodicity 51
  - 3.3.1 Composition Rules 51
  - 3.3.2 Aperiodicity of Robinson Triangles 54
  - 3.3.3 Local Properties 55
- 3.4 Index Sequences 59
  - 3.4.1 Some Periodic Index Sequences 60
  - 3.4.2 Classification of Tilings with Robinson Triangles 66
- 3.5 On the Density of Large and Small Triangles 74
- 3.6 Final Comments 82

4 Penrose Tilings 85
- 4.1 Aperiodicity, Classification, Local Properties, Symmetries 90
- 4.2 Imperfect Substitution Rules 95
- 4.3 Conway Worms 98

4.3.1 Musical Sequences ... 101
4.3.2 Hierarchical Structure ... 103
4.3.3 Fibonacci Strings ... 104
4.3.4 Fibonacci Tilings ... 106
4.4 Ribbons ... 108
4.5 Non-locality and Empires ... 114
4.6 The Three-Color Theorem ... 117

5 De Bruijn's Pentagrids ... 121
5.1 Pentagrids ... 122
5.2 The Cut-and-project Method ... 133
5.2.1 One-dimensional Tilings ... 133
5.2.2 Penrose Rhombi ... 139
5.3 Composition and Pentagrids ... 141
5.4 Congruence Classes of Tilings ... 151
5.5 Singular Pentagrids ... 153

6 The Noncommutative Space of Penrose Tilings ... 157
6.1 Topology of the Space of Penrose Tilings ... 159
6.1.1 The Cantor Set ... 159
6.1.2 The Space of Penrose Tilings ... 160
6.2 The Canonical Anticommutation Relations ... 162
6.3 Approximately Finite-Dimensional C*-algebra ... 166
6.3.1 Morphisms of Finite-Dimensional C*-Algebras ... 166
6.3.2 Bratteli Diagrams ... 168
6.3.3 K-Theory of AF Algebras ... 169
6.4 AF Equivalence Relations ... 174
6.4.1 Étale Relations and Their Convolution Algebra ... 177
6.4.2 Cantor Spaces of Infinite Paths ... 179
6.4.3 From Bratteli Diagrams to AF Relations ... 184
6.4.4 From AF Relations Back to AF Algebras ... 187

Appendix A: Some Useful Formulas ... 191

References ... 197

Chapter 1
Introduction

The problem of covering a flat surface—a subset of the Euclidean plane or the whole plane itself—using some fixed geometric shapes and with no overlaps is probably one of the oldest in mathematics. Such a covering is called a *tiling*, or also a *tessellation* or a *mosaic* (see Definition 2.1). The one in Fig. 1.1, for example, is a tiling with equilateral triangles and (non-regular) pentagons.

Mosaics with colored geometrical shapes can be seen in stained glass windows in Christian churches. Tilings using shapes that are invariant under a rotation of 72°, such as regular pentagons or pentagrams, occur frequently in Islamic art. These shapes are said to possess a 5-fold symmetry (since 72 is one-fifth of 360).

Probably, one of the reasons for the popularity of this topic is that many of its fundamental problems are simple to formulate, even if they may require advanced tools to be solved. Not always, though; the story of the mathematical amateur Marjorie Rice, illustrated e.g. in [Sch78], is a famous example of a non-professional mathematician making an important original contribution to the topic.

Despite the apparently random placement of tiles in Fig. 1.1, an optical effect due to the fancy choice of colors, it is not difficult to realize that one can extend the tiling to the whole plane simply by translating copies of the same basic configuration

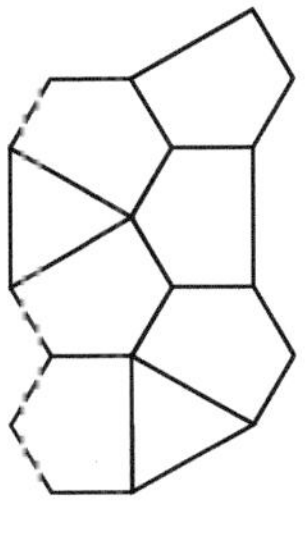

In this way, one gets a first, albeit not the easiest, example of *periodic* tiling, wherein the same pattern is repeated over and over again.

F. D'Andrea, *A Guide to Penrose Tilings*,
https://doi.org/10.1007/978-3-031-28428-1_1

Fig. 1.1 A portion of a tiling of the plane

In 1961, Wang proposed the following problem [Wan61]. Consider a finite number of square tiles, with edges marked by symbols (or colors, letters, numbers, ...). We want to tile the whole plane by placing copies of these tiles, which we call *Wang tiles*, in the plane. The tiles can be translated but not rotated, and two adjacent tiles must share a full edge: this is what we call an *edge-to-edge* tiling. Moreover, there is a *matching rule*: two edges can touch only if they are decorated with the same symbol. The Domino Problem asks if there exists an algorithm that can decide whether a set of Wang tiles can tile the whole plane. Wang proved that the Domino Problem is decidable if and only if there does not exist an *aperiodic* set of Wang tiles, i.e. a set that can tile the plane only non-periodically, and for this reason he conjectured that such a set didn't exist. The first aperiodic set of Wang tiles was found by Berger in 1966 [Ber66] and consisted of more than 20.000 different tiles (but Berger's original Ph.D. Thesis contained a simplified version with 104 tiles). A notable contribution to this topic is the one by Knuth, the creator of TEX, who found an aperiodic set of 92 Wang tiles [Knu69, Sect. 2.3.4.3]. Many others contributed to the subject (we will not discuss the Domino Problem in this book, but the interested reader can find more about it in [GS87]). The number of Wang tiles in an aperiodic set was recently reduced to 11, and it was proved that this is the minimal number [JR15].

The story of the Domino Problem intertwines with another one: the search for tilings of the plane that use only shapes with 5-fold symmetry. It is well known that this cannot be achieved with a single tile [DGS82], for example a regular pentagon, or by repeating the same pattern over and over again (see Sect. 2.3). This problem can be found already in Kepler's treatise *Harmonices mundi* (1619), where one can find several mosaics with regular pentagons, decagons, and pentagrams. Of course, one can tile the whole plane with a single pentagonal tile if the angles are chosen correctly. For example, this can be achieved with the pentagons in Fig. 1.1 by repeating the pattern

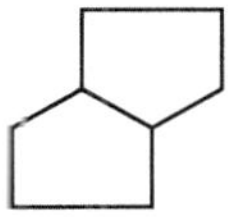

over and over again. Kepler proved that, up to a similarity, there exist exactly eleven different edge-to-edge tilings of the plane with convex regular polygons that are semiregular, i.e. that have only one type of vertex neighborhood (see Sect. 2.1 for the definition). Three of them are made with a single polygon (i.e. they are *monohedral*): either an equilateral triangle, a square, or a regular hexagon.

In 1973, looking for tilings of the plane with shapes with 5-fold symmetry, Penrose found a set of 6 tiles that can only tile the plane non-periodically [Pen74], thus reducing the cardinality of known aperiodic sets to 6. One year later, he reduced this number to 2, by discovering his famous tiles shaped like a *kite* and a *dart* (see Chap. 4 for the details):

Even if the discovery was in 1973–74, the first official mention was only in 1977 in [Gar77] (as explained therein, Penrose applied for patents in the UK, US and Japan, and was reluctant to advertise the tilings before the patents were granted). The paper [Pen78] contains an account of how Penrose discovered his sets with 6 and 2 tiles, and a nice Escheresque non-periodic tiling with birds (now called "Penrose non-periodic chickens"): it was produced using the same technique employed by Escher to transform polygonal tiles into animal shapes.

Let us mention that Socolar and Taylor recently found an aperiodic disconnected monotile [ST11, ST12]. Even more recently, in the preprint [SMKGS23], the Authors exhibit a tile homeomorphic to a disk that, when used together with its reflection, can tile the plane only non-periodically (the interesting story of this discovery can be found, e.g., in [Kla23]). An overview of previously known aperiodic sets of tiles can be found in Chap. 7 of [Sen95].

There exists many books on Penrose tilings intended for a general audience (e.g. [Gar97]), survey papers (e.g. the original papers by Penrose [Pen74, Pen78]) and even some excellent YouTube videos (e.g. [Mul20]) carefully explaining all the properties of these tilings.

An efficient way to study Penrose tilings is by dissecting them into triangles:

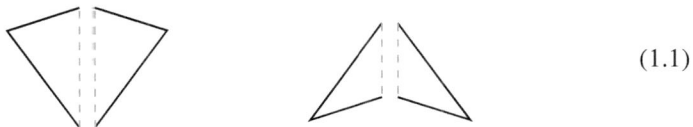

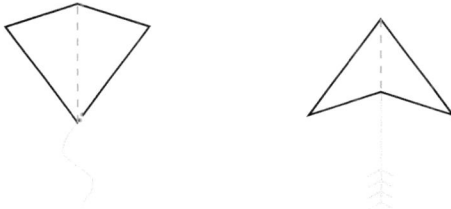

(1.1)

This was the method used by Robinson in [Rob75], that we illustrate in Chap. 3. Robinson triangles are extremely useful to study local properties of Penrose tilings. Moreover, they allow to introduce a fundamental tool in the classification of Penrose tilings up to isometries: the *index sequence* of a tiling (Sect. 3.4).

A few years later, another fundamental tool for the construction of Penrose tilings of the plane was discovered by de Bruijn: the so-called *pentagrid* method [dB81]. A very nice historical account of the impact of de Bruijn on this subject is in [AYP13]. Rather than tiling the plane by constructing bigger and bigger patches, de Bruijn discovered a global construction using five bundles of parallel lines. A byproduct of this method is the interpretation of Penrose non-periodic tilings as two-dimensional slices of periodic tilings of a five-dimensional space. At the core of de Bruijn's idea is the observation that the projection

$$\mathbb{Z}^n \to \mathbb{C}, \qquad (k_0, \dots, k_{n-1}) \mapsto \sum_{j=0}^{n-1} k_j (e^{2\pi \mathrm{i}/n})^j, \qquad (n \geq 1)$$

of an n-dimensional lattice on a plane is a lattice if $n \in \{1, 2, 3, 4, 6\}$, and has dense image in $\mathbb{C}$ in all other cases. For $n = 5$, if we project only points that are "near" the plane, we get a rhombus tiling that is equivalent (in a suitable sense) to a Penrose tiling by kites and darts. The image of the above projection is the ring of cyclotomic integers $\mathbb{Z}[e^{2\pi \mathrm{i}/n}]$, and the pentagrid method is a bridge between Penrose tilings and algebraic number theory.

An important stimulus for the study of non-periodic tilings came from a discovery in chemistry. In 1982, Dan Shechtman observed a diffraction pattern with 10-fold symmetry from a metal alloy sample. Such a symmetry is impossible in ordinary crystals, that are modeled on lattices (cf. Sect. 2.3). For his work, Shechtman was awarded the Nobel Prize in Chemistry in 2011 for the discovery of quasi-crystals [Fer18].

A (doubly) periodic tiling is formed by repeating a bounded region infinitely many times. It follows that every finite subset of tiles (every "patch" in the terminology that we will introduce in the next chapter) is repeated infinitely many times in the tiling. A tiling with the latter property is called *repetitive*, and surprisingly a tiling does not need to be periodic to be repetitive. Penrose tilings, for example, are non-periodic and repetitive. A tiling that is non-periodic and repetitive is called *quasi-periodic*. Quasi-periodic tilings have attracted the interest of many scientists because they provide a mathematical model for quasi-crystals.

Using index sequences one shows that the space parameterizing inequivalent Penrose tilings is the quotient of a Cantor space by an equivalence relation with dense equivalence classes. The quotient topology is then the trivial (indiscrete) one, and gives no insight on the structure of the space of Penrose tilings. In the spirit of Noncommutative Geometry [Con94], quotient spaces that are not Hausdorff can be studied using groupoid C*-algebras, and the space of equivalence classes of Penrose tilings is an example of such a "bad" quotient space. It is therefore natural to use C*-algebra techniques to learn more about these tilings. The use of C*-algebra techniques in solid state physics was popularized by Bellissard in the '80s [Bel86]. A good starting point for their use in the study of tilings is the beautiful survey by Kellendonk and Putnam [KP00] (see also [Sad08]).

This book is structured as follows. The second chapter is a general introduction to tilings and illustrates the main notions through some simple examples. The third chapter is about Robinson's triangles. The fourth chapter is about Penrose tilings. The fifth chapter is about de Bruijn's pentagrid method, and the realization of a Penrose tiling as a projection of a portion of a five-dimensional lattice. The sixth and last chapter is about the noncommutative geometry of Penrose tilings. The geometric properties Robinson triangles and Penrose tiles are intimately related to the algebraic properties of the *golden ratio*. Some of these algebraic properties are collected in Appendices A.1 and A.2.

We shall adopt the following notations. We denote by $\mathbb{N}$ the set of natural numbers including 0. If S is a set, $|S|$ denotes its cardinality. If S is a subset of a topological space, $\overline{S}$ is its closure and $\mathring{S}$ its interior. $A \subseteq B$ or $B \supseteq A$ means that A is a subset of B, while $A \subset B$ or $B \supset A$ means that A is a proper subset of B. An open interval with endpoints a and b, is denoted by $]a, b[$, a closed one by $[a, b]$.

Chapter 2
Tilings and Puzzles

2.1 Preliminary Notions

In the whole book we will identify $\mathbb{R}^2$ with $\mathbb{C}$ in the usual way, and we will refer to it as the (Cartesian) plane. We denote by

$$D_r(z_0) := \{z \in \mathbb{C} : |z - z_0| < r\}$$

the open disk of radius $r > 0$ centered at $z_0 \in \mathbb{C}$.

If $f : \mathbb{C} \to \mathbb{C}$ is a map and $X \subseteq \mathbb{C}$ a set, we define $f(X) := \{f(x) : x \in X\}$. If $\mathcal{S} = \{S_\alpha\}$ is a family of subsets of $\mathbb{C}$ (labeled by some index α), we define $f(\mathcal{S}) := \{f(S_\alpha)\}$. A set X is invariant under f if $f(X) = X$ (which does not mean that $f(x) = x \ \forall\, x \in X$), and a family is invariant under f if $f(\mathcal{S}) = \mathcal{S}$ (which does not mean that $f(S_\alpha) = S_\alpha \ \forall\, \alpha$). A *symmetry* of $\mathcal{S}$ is a bijective map $f : \mathbb{C} \to \mathbb{C}$ such that $f(\mathcal{S}) = \mathcal{S}$. The identity transformation Id, defined by $\mathrm{Id}(z) = z \ \forall\, z \in \mathbb{C}$, is a symmetry of any family $\mathcal{S}$.

Observe that, if f is a symmetry of $\mathcal{S}$, applying f^{-1} to both sides of the identity $f(\mathcal{S}) = \mathcal{S}$ we find that f^{-1} is a symmetry as well. If f and g are symmetries of $\mathcal{S}$, the composition $f \circ g$ is a symmetry as well. The set of all symmetries of $\mathcal{S}$ is then a subgroup of the group of bijections $\mathbb{C} \to \mathbb{C}$, with operation given by the composition and neutral element given by the identity map. A symmetry is *non-trivial* if it is not the identity transformation.

We are interested in symmetries that are rigid motions of the plane: rotations, translations and reflections. A *translation*, in complex coordinates, has the form

$$f(z) = z + t$$

for some $t \in \mathbb{C}$. A *rotation*, of an angle θ and around a point $c \in \mathbb{C}$, has the form

$$f(z) = u(z - c) + c$$

F. D'Andrea, *A Guide to Penrose Tilings*,
https://doi.org/10.1007/978-3-031-28428-1_2

with $u = e^{i\theta}$. A *scaling* with scale factor $\lambda > 0$ is a transformation of the form $f(z) = \lambda z$. Scalings, translations and rotations generate the group $\mathrm{Aff}(\mathbb{C})$ of affine transformations of the (complex) plane, that are of the form $z \mapsto az + b$ for some $(a, b) \in \mathbb{C}^\times \times \mathbb{C}$. (Here $\mathbb{C}^\times := \mathbb{C} \setminus \{0\}$.) Rotations and translations alone generate the subgroup of *direct* isometries, those transformations of the form $z \mapsto uz + b$ with $b \in \mathbb{C}$ and $u \in \mathbb{S}^1 := \{z \in \mathbb{C} : |z| = 1\}$. Finally, a *reflection* across a line ℓ is a transformation of the form

$$f(z) = z - 2u\,\mathrm{Re}\big(\bar{u}(z - z_0)\big),$$

where u is a unit vector orthogonal to ℓ and $z_0 \in \ell$ any point. Reflections and direct isometries generate the group of *isometries* of the plane. Reflections and affine transformations generate the group of *similarities* of the plane (Fig. 2.1).

Two subsets X and Y of $\mathbb{C}$ (resp. two families $\mathcal{S}$ and $\mathcal{S}'$) are called *similar* if there exists a similarity f such that $f(X) = Y$ (resp. $f(\mathcal{S}) = \mathcal{S}'$), and they are called *congruent* if there exists an isometry f such that $f(X) = Y$ (resp. $f(\mathcal{S}) = \mathcal{S}'$). Both similarity and congruence are equivalence relations, since they come from group

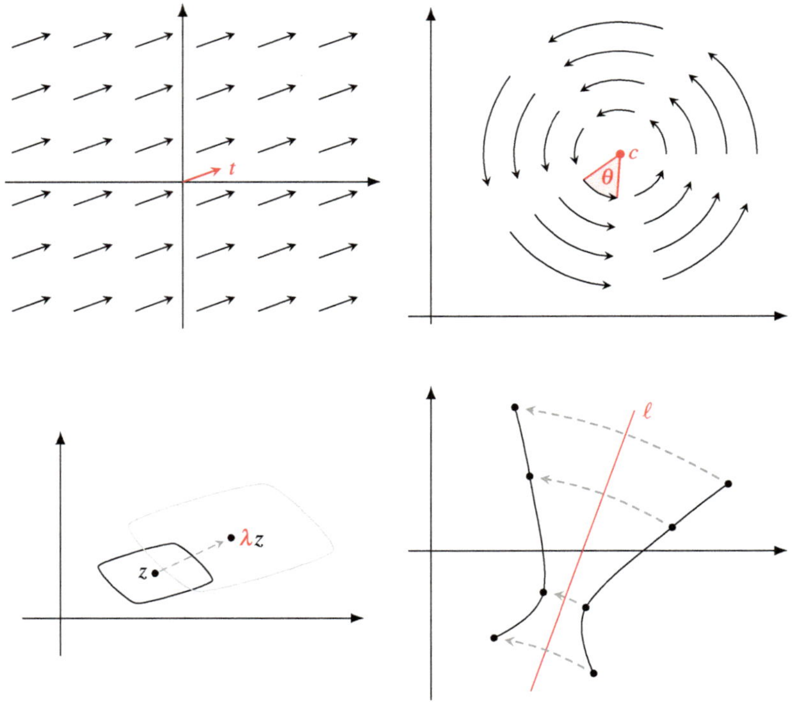

Fig. 2.1 Translations, rotations, scalings and reflections

actions. Finally, for lack of a better name, X and Y (resp. $\mathcal{S}$ and $\mathcal{S}'$) will be called *equivalent* if there exists a direct isometry f such that $f(X) = Y$ (resp. $f(\mathcal{S}) = \mathcal{S}'$).

If $(a, b) \in \mathbb{C}^\times \times \mathbb{C}$, $X \subseteq \mathbb{C}$ and $\mathcal{S} = \{S_\alpha\}$ is a family of subsets of $\mathbb{C}$, consistently with the notations introduced above we define

$$aX + b := \{ax + b : x \in X\},$$
$$a\mathcal{S} + b := \{aS_\alpha + b\}.$$

The equivalence class of $T \subseteq \mathbb{C}$ (under the action of direct isometries) is

$$[T] := \{uT + b : (u, b) \in \mathbb{S}^1 \times \mathbb{C}\}.$$

Definition 2.1 (*Tilings*)

(i) A *tile* is a closed subset of $\mathbb{C}$ with non-empty interior. The equivalence class of a tile (under the action of direct isometries) is called a *prototile*.

(ii) A *partial tiling* (or *packing*) is a family $\mathcal{T} = \{T_\alpha\}$ of tiles with non-overlapping interiors:

$$\mathring{T}_\alpha \cap \mathring{T}_\beta = \varnothing \ \forall \, \alpha \neq \beta.$$

The set of prototiles of a partial tiling is called its *protoset*.

We will say that $\mathcal{T}$ *covers* a subset X of $\mathbb{C}$ if:

$$X \subseteq \bigcup_\alpha T_\alpha.$$

When the equality holds, we say that $\mathcal{T}$ covers *exactly* X (or that *tiles* X). If further $X = \mathbb{C}$, we call $\mathcal{T}$ a *tiling of the plane*, or a *tiling* 'tout court'.

(iii) A partial tiling with protoset $\mathfrak{P}$ is called *legal* if it can be extended to a tiling (of the whole plane) with the same protoset $\mathfrak{P}$.

(iv) A finite subset $\mathcal{P} \subseteq \mathcal{T}$ of a tiling is called a *patch*. Thus, a patch is a finite legal partial tiling.

One can verify that a similarity transforms a tiling into another tiling (without changing the protoset if the similarity is a direct isometry).

It is straightforward to generalize the notion of tiling to an arbitrary topological space. In this book, however, we are only interested in tilings of the plane (with small exceptions in Sects. 4.3.4 and 5.2.1).

Definition 2.2 (*Symmetries*)

(i) A tiling $\mathcal{T}$ is called *non-periodic* if, for all $b \in \mathbb{C}$, $\mathcal{T} + b = \mathcal{T}$ implies $b = 0$. It is called *periodic* if it is not non-periodic, and *bi-periodic* if it is invariant under translations in the direction of two linearly independent vectors.

(ii) A family $\mathfrak{P}$ of prototiles is *aperiodic* if there exist tilings (of the whole plane) with protoset $\mathfrak{P}$ and they are all non-periodic.

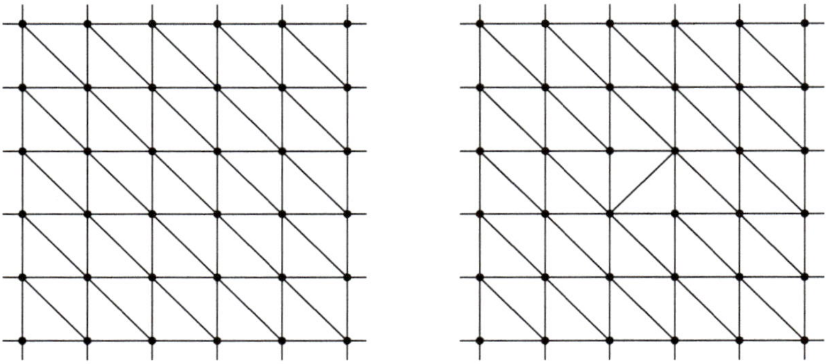

Fig. 2.2 Bi-periodic and non-periodic monohedral tilings

Fig. 2.3 Edge and vertex neighbourhoods

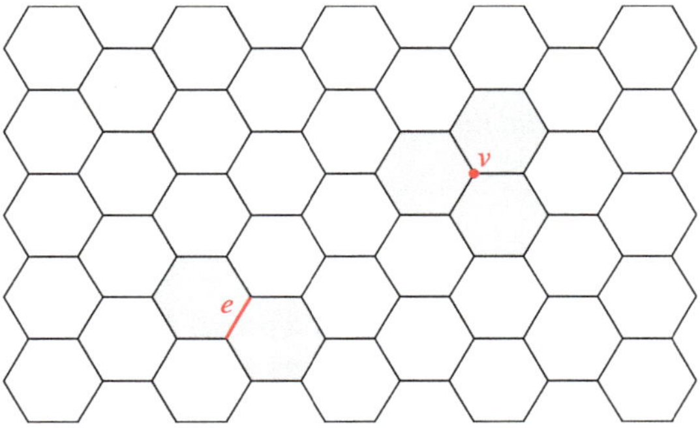

A (partial) tiling is called *n-hedral* if the set of prototiles is finite with n elements (*monohedral* if $n = 1$, *dihedral* if $n = 2$, etc.).

A basic example of monohedral tiling with unit squares is obtained by placing the vertices on points with integer coordinates. If we divide each square in two parts along the descending diagonal, we get a bi-periodic tiling of the plane by triangles, as illustrated in the first picture in Fig. 2.2. If we rotate by 90 degrees a single square, we obtain an example of monohedric non-periodic tiling (cf. Fig. 2.2).

A (partial) tiling is called *edge-to-edge* if its tiles are polygons and, whenever the intersection $T \cap T'$ of two tiles intersects the interior of an edge (of either of the two tiles), it contains the full edge. For example, the tilings in Fig. 2.2 are edge-to-edge. Given an edge-to-edge (partial) tiling: two tiles are called *neighbors* if their intersection is non-empty; they are called *adjacent* if they share a (full) edge; a pair of tiles sharing an edge e is called *edge neighbourhood* of e; the set of all tiles containing a vertex v is called *vertex neighbourhood* of v. An example of edge and vertex neighbourhoods in a "honeycomb" tiling is in Fig. 2.3.

In Fig. 2.4 we see two examples of tilings with polygons (squares) that are not edge-to edge, one monohedral (all squares are of the same size) and one dihedral (it uses two sizes of square tiles). It is not difficult to imagine how these patterns extend to the whole plane.

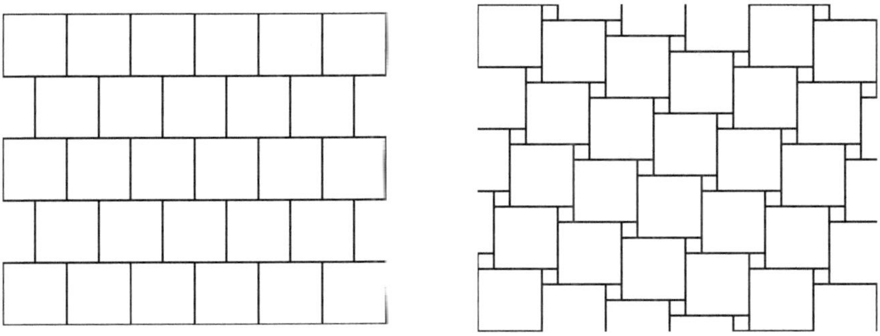

Fig. 2.4 Polygonal tilings that are not edge-to-edge

One last important notion that we need to mention is the one of repetitivity. If $\mathcal{T}$ is a tiling and $\mathcal{P} \subset \mathcal{T}$ a patch, by a *copy* of $\mathcal{P}$ we mean another patch $\mathcal{P}'$ (in the same or in a different tiling) such that $\mathcal{P}' = u\mathcal{P} + b$, with $(u, b) \in \mathbb{S}^1 \times \mathbb{C}$. We will call it a *translated copy* if $\mathcal{P}' = \mathcal{P} + b$. We say that $\mathcal{P}$ and $\mathcal{P}'$ have the same *orientation* if one is a translated copy of the other, and that they have different orientations otherwise.

Definition 2.3 A tiling $\mathcal{T}$ is called *repetitive* if, for every patch $\mathcal{P}$, the tiling contains infinitely many (distinct) translated copies of $\mathcal{P}$.

Periodic tilings are obviously repetitive. But this is not the only class of examples.

Definition 2.4 A tiling is called *quasi-periodic* if it is non-periodic and repetitive.

We will see that Penrose tilings are quasi-periodic (Proposition 4.23).

2.2 Square Tilings, Puzzles and Matching Rules

Before moving on with more complicated examples, let us illustrate some basic ideas using some monohedral tilings that are minimal modifications of tilings with unit squares. Consider the following (equivalence classes of) puzzle pieces:

$[T_0] :=$ 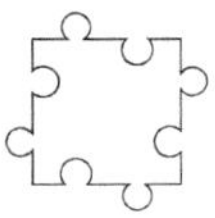 , $[T_1] :=$ 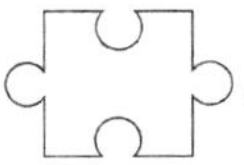 , $[T_2] :=$  ,

where T_0, T_1, T_2 are some chosen representatives. Any tiling of the whole plane with one of these prototiles is necessarily "edge-to-edge": the only way to cover the white space inside a jigsaw piece is by interlocking another piece, e.g.

(2.1)

The piece T_0 is desined so that any edge can be attached to any other edge, and it is invariant under 90° rotations around its center.[1] From any edge-to-edge tiling of the plane with unit squares, we can then obtain a tiling with prototile $[T_0]$ by replacing each square with a copy of T_0, and vice versa.

We can build a tiling of the whole plane with prototiles $[T_1]$ or $[T_2]$ using the following trick. First, construct the two building blocks (one using only T_1 and one using only T_2 as prototiles):

(2.2)

Then, observe that both building blocks can fit in a tile T_0 if we double its size. Now it is clear how to obtain a tiling of the whole plane: we scale any T_0-tiling of the plane by a factor two, and then put inside each tile one of the blocks in (2.2). If we only use one of the two building blocks, we get monohedral tilings. But we can use both as well, and get a dihedral tilings with protoset $\{[T_1], [T_2]\}$.

We saw that every T_0 tiling of the plane is equivalent to an edge-to-edge tiling by unit squares. What about the other two prototiles? They have two different types of edges, one with a hole and one with a knob, and two edges can be attached only if they have different type.

Since drawing squiggly edges (especially by hand) is complicated, it is often more convenient to put markings on the edges, which tell us how the tiles must fit together. For example, one can decorate the prototiles with arrows:

$[\overline{T}_1] :=$ 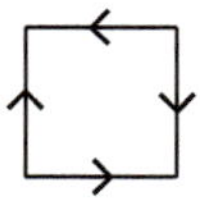 $[\overline{T}_2] :=$ 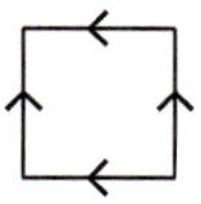

and allow only edge-to-edge tilings where adjacent tiles have overlapping arrows pointing in the same direction. This is a first example of a *matching rule*.

It is not difficult to convince oneself that the protosets $\{[T_1]\}$ and $\{[\overline{T}_1]\}$ are "interchangeable", in the sense that any T_1-tiling of the whole plane can be transformed into

[1] Here we are a bit sloppy with the use of words such as "edge" and "center", since the prototiles are neither squares nor polygons, but the meaning should be clear.

a $\overline{T}_1$-tiling satisfying the matching rule, and vice versa. Analogously, the protosets $\{[T_2]\}$ and $\{[\overline{T}_2]\}$ are interchangeable.

Now, what about symmetries? An edge-to-edge tiling of the plane by unit squares (or with prototile T_0) is obviously invariant if we move all squares one position up or to the right. As a consequence, T_1- and T_2-tilings constructed with the building blocks (2.2) with the inflation and substitution procedure described above are always invariant if we shift every tile by 2 positions (in any direction). A shift by 1 position, on the other hand, will move a tile over one with different orientation and will not be a symmetry of the tiling.

It is not difficult to produce a T_2-tiling that is invariant under shifts by 1. A portion of such a tiling will look like:

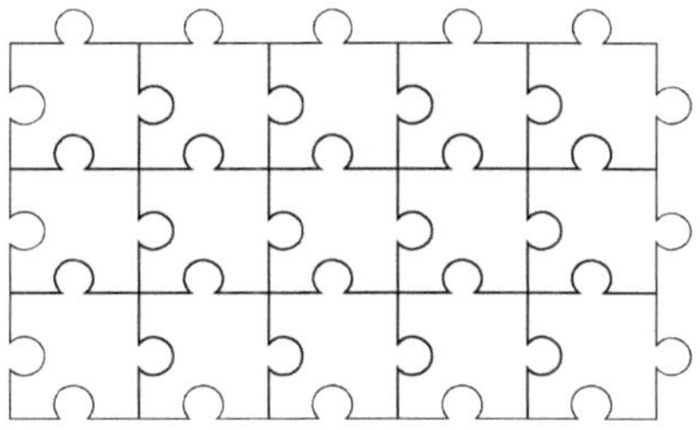

(2.3)

This is not possible with the T_1 tile: two adjacent tiles must have different orientation, otherwise they don't interlock.

Definition 2.5 We say that a tiling $\mathcal{T}$ has an *n-fold* symmetry around a point $c \in \mathbb{C}$ if it has a rotational symmetry around c and the group of rotations around c that are symmetries of $\mathcal{T}$ is a cyclic group of order n, i.e. it is generated by the rotation $z \mapsto e^{2\pi \mathrm{i}/n}(z - c) + c$.

The standard tiling of the plane with squares has 4-fold symmetry: the tiling is invariant under a rotation of 90° around any vertex. What if the center c of rotation is not a vertex? If c is internal to an edge, the rotation should map the edge to itself, which means that c must be a middle-point and the angle 180°. If c is in the interior of a square, the rotation should map the square to itself, which forces c to be the center and the angle to be 90°.

Now, if one tries to rotate around the central vertex by multiples of 90° the building blocks in (2.2), forgetting about the colors that are there only for visual reasons, one soon realizes that this rotation is a symmetry of the blocks. It follows that T_1- and T_2-tilings that are constructed from a T_0-tiling as explained above have 4-fold symmetry around vertices that are at the center of building blocks; the T_1-tiling has an additional 2-fold symmetry around the center of any tile, and an argument similar to the one given for squares shows that both the T_1- and T_2-tiling have no other non-trivial rotational symmetries besides the ones just described. The T_2-tiling in (2.3), on the other hand, has no non-trivial rotational symmetries, since all the tiles have the wedge between two knobs pointing in the same direction, and a non-trivial rotation around any point would change such a direction.

Consistently with the terminology adopted for families of sets, we give the following definition.

Definition 2.6 Two (partial) tilings are called *equivalent* if they can be transformed one into the other with a direct isometry. (It is obvious that equivalent tilings must have the same protoset.)

For example, the two partial tilings in (2.1) are not equivalent: there is no way to transform one into the other with a rigid motion.

This notion of equivalence is particularly interesting for tilings of the whole plane, since we can only draw portions of them and it is usually not possible to find by trial and error if one tiling can be transformed into another. One has to develop suitable methods to find out in how many inequivalent ways it is possible to tile the plane with a fixed set of prototiles. As a warm up exercise, let us study the three protosets $\{[T_0]\}$, $\{[T_1]\}$ and $\{[T_2]\}$.

Two edge-to-edge tilings of the plane by unit squares (hence by T_0) are always equivalent: any isometry transforming one square of the first tiling into one of the second tiling will do the job. Although less obvious, there is a unique T_1-tiling (up to equivalence) as well: call a tile congruent to T_1 "blue" if the edges with a hole are parallel to the real axis, and "magenta" if they are parallel to the imaginary axis. Now, given any tiling of the plane, with a direct isometry we can transform it into one whose vertices have integer coordinates (thus: they form the lattice $\mathbb{Z}^2$). Let us call *fundamental domain* the square with vertices $(0, 0)$, $(0, 1)$, $(1, 0)$ and $(1, 1)$. With a translation we can always put a blue tile in the fundamental domain, and since the colors of the tiles must be alternating, by induction one proves that if the bottom-left vertex of a tile has coordinates with the same parity, then it must be blue, otherwise it is magenta, like in the following picture:

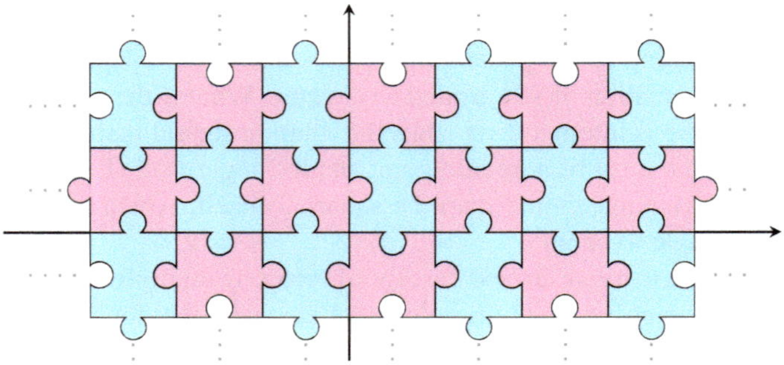

Thus, there is a unique T_1-tiling up to rotations and translations.

Finally, we pass to the prototile T_2. Firstly, observe that equivalent tilings must have the same symmetries. Thus, there exist at least two are inequivalent T_2-tilings: one with 4-fold symmetry constructed with the inflation and substitution procedure, and the tiling with no non-trivial rotational symmetries in (2.3). We now show that in fact there are infinitely many.

To this aim, it is convenient to decorate the prototile with arrows:

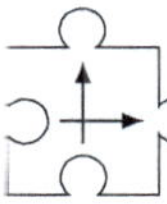

We can now forget about the shape of the edges and simply consider edge-to-edge tilings with prototile

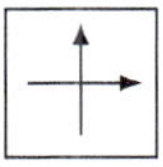

with the matching rule that consecutive arrows should point in the same direction.

For example:

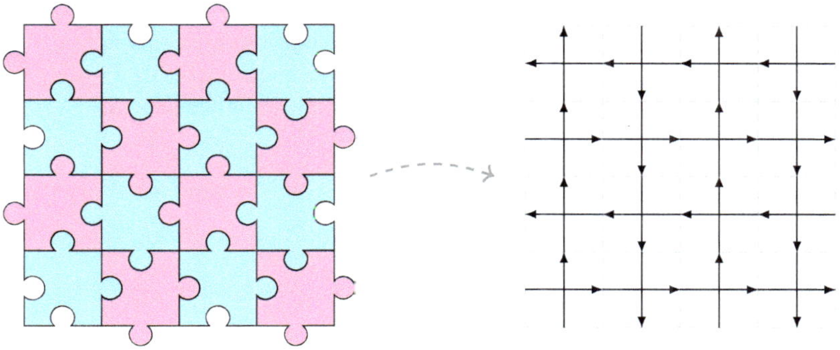

Given any T_2-tiling of the plane, we can rotate and translate it so that the center of each tile has integer coordinates. Then, a T_2-tiling of the plane will be simply a collection of arrows—one for each edge of the lattice $\mathbb{Z}^2$—such that consecutive arrows point in the same direction. By induction, in every horizontal or vertical line of the lattice, all arrows should point in the same direction, and the direction of every line can be chosen independently from the others. For example, given a horizontal line with arrows pointing right, by rotating all the tiles in that line we can change the orientation of all horizontal arrows without changing the orientation of the vertical ones:

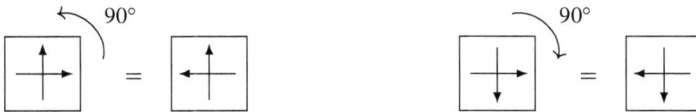

Thus, we can always change the direction of a line in a T_2-tiling without breaking the matching rule. We can encode the direction of lines using two binary sequences

$$h := (\ldots, h_{-2}, h_{-1}, h_0, h_1, h_2, \ldots) \qquad v := (\ldots, v_{-2}, v_{-1}, v_0, v_1, v_2, \ldots)$$

by setting $h_i = 0$ if the horizontal line with ordinate i has arrows pointing left and $h_i = 1$ otherwise, $v_i = 0$ if the vertical line with abscissa i has arrows pointing down and $v_i = 1$ otherwise. In this way, we get a correspondence between equivalence classes of T_2-tilings of the plane and pairs of binary sequences. We can now prove that there are infinitely many equivalence classes of T_2-tilings of the plane.

Proposition 2.7 *The set of equivalence classes of T_2-tilings has cardinality $|\mathbb{R}|$.*

Proof Let X be the set of all binary sequences $(a_n)_{n\in\mathbb{Z}}$ with $a_0 = 1$, and consider the equivalence relation on X, given by:

$$(a_n)_{n\in\mathbb{Z}} \sim (b_n)_{n\in\mathbb{Z}} \quad \Longleftrightarrow \quad \exists\, i_0 \in \mathbb{Z} : a_k = b_{k+i_0}\ \forall\, k \in \mathbb{Z}.$$

Given any T_2-tiling, there is exactly one rotation that transforms the tile around the origin into one with arrows pointing up and right, i.e. into one whose associated pair of binary sequences has $h_0 = v_0 = 1$. Now, two such tilings are equivalent if and only if they are related by a translation. The set of equivalence classes of T_2-tilings of the plane has then the same cardinality of the set $X/{\sim}$ above (since $|2 \times X/{\sim}| = |X/{\sim}|$ if $|X/{\sim}| = |\mathbb{R}|$). Clearly

$$|X/{\sim}| \leq |X| \leq |2^{\mathbb{Z}}| = |\mathbb{R}|.$$

Let

$$Y := \big\{(a_n)_{n\in\mathbb{Z}} \in X : a_k = 0\ \forall\, k < 0\big\}.$$

Observe that the restriction to Y of the quotient map $X \to X/{\sim}$ is injective, since shifting a sequence in Y to the right gives a new sequence $(a_n)_{n\in\mathbb{Z}}$ with $a_0 = 0$ (hence not belonging to X), and shifting it to the left gives a new sequence $(a_n)_{n\in\mathbb{Z}}$ with $a_k = 1$ for at least one $k < 0$ (hence not belonging to Y). Thus, $|Y| \leq |X/{\sim}|$.

Finally, an injective map from $2^{\mathbb{Z}}$ to Y is given by

$$(a_n)_{n\in\mathbb{Z}} \mapsto (\ldots, 0, \ldots, 0, 1, a_0, a_1, a_{-1}, a_2, a_{-2}, \ldots).$$

Therefore, $|\mathbb{R}| = |2^{\mathbb{Z}}| \leq |Y| \leq |X/{\sim}|$, proving that $|X/{\sim}| = |\mathbb{R}|$. ■

2.3 Symmetries of Tilings

We already met examples of tilings with 4-fold symmetry. Monohedral tilings with 2-, 3-, and 6-fold symmetry can be easily produced using non-square rectangles, regular hexagons, and equilateral triangles, respectively. The crystallographic restriction theorem states that these are the only possible symmetries of lattices (that we use to model crystals). They are also the only possible symmetries of edge-to-edge monohedral tilings with regular polygons. This is the content of the next proposition.

Proposition 2.8 *If $\mathcal{T}$ is an edge-to-edge monohedral tiling with prototile given by a regular polygon and $z \mapsto e^{2\pi i\theta}(z - c) + c$ is a rotational symmetry, then it must be $\theta \in \left\{\frac{1}{3}, \frac{1}{4}, \frac{1}{6}\right\}$ (3-fold, 4-fold and 6-fold symmetry, respectively).*

Proof Let the prototile be a regular n-gon ($n \geq 3$). One has $k \geq 2$ polygons meeting on a vertex. Since the interior angle of each n-gon at a vertex is $\frac{n-2}{n}\pi$, it should be

$$\frac{2\pi}{k} = \frac{n-2}{n}\pi, \qquad \text{that is } k = \frac{2n}{n-2} = 2 + \frac{4}{n-2}.$$

This is integer if and only if $n \in \{3, 4, 6\}$, since $0 < \frac{4}{n-2} < 1$ if $n \geq 7$. ∎

If we abandon regular polygons, things get more complicated. Using Euler's formula for planar graphs it can be shown that in an edge-to-edge monohedral tiling of the plane with a convex polygon, the number of edges of the prototile is at most six. It is then reasonable to believe that there are only finitely many convex polygons one can tile the plane with. The problem however is still open [Zon20]. Rao completed recently a classification of pentagonal tilings [Rao17].

Open problem 2.9 *Classify all the convex polygons that may appear in monohedral tilings of the plane.*

Recall that a *lattice* in $\mathbb{C}$ is a subset of the form $\mathbb{Z}v_1 + \mathbb{Z}v_2$, where $v_1, v_2 \in \mathbb{C}$ are linearly independent over $\mathbb{R}$ (an illustration is in Fig. 2.5). A tiling is bi-periodic if and only if its symmetry group contains a lattice (and a periodic tiling with a non-trivial rotational symmetry is bi-periodic, cf. Lemma 2.10).

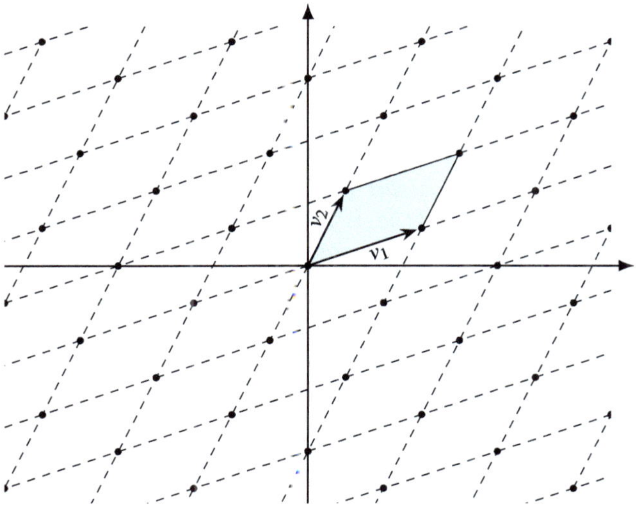

Fig. 2.5 A lattice in $\mathbb{R}^2$

Observe that, for an edge-to-edge tiling whose vertex set is a lattice, not every symmetry of the lattice is a symmetry of the tiling (the latter should send not only vertices to vertices but edges to edges as well). For example, the first tiling in Fig. 2.2 has 2-fold symmetry, even if the underlying lattice has 4-fold symmetry. The second tiling in Fig. 2.2 is an example where the tiling has no translational symmetry, even if its vertex set is a lattice. Moreover, the vertex set of an edge-to-edge tiling, even a bi-periodic one, is not necessarily a lattice (it could be a union of translated copies of a lattice, for example).

We aim at proving that, under reasonable assumptions, the only rotational symmetries of a periodic tiling are 2-, 3-, 4- and 6-fold. We stress that this result cannot be derived from the crystallographic restriction theorem, and must be proved independently. We start with some elementary lemmas that are of independent interest.

Lemma 2.10 *(i) A non-periodic tiling cannot have non-trivial rotational symmetries around two distinct points.*
(ii) If a periodic tiling has a non-trivial rotational symmetry of an angle $\theta \notin \pi\mathbb{Z}$, then it is bi-periodic.

Proof (i) Let $\alpha : z \mapsto u(z-c)+c$ and $\alpha' : z \mapsto u'(z-c')+c'$ be two rotations and observe that $\alpha^{-1} : z \mapsto u^*(z-c)+c$. The composition

$$t(z) := \alpha'\alpha\alpha'^{-1}\alpha^{-1}(z) = z + (u-1)(u'-1)(c'-c)$$

is a translation. If α and α' are symmetries of a tiling, then t is a symmetry as well and it is non-trivial if $u \neq 1$, $u' \neq 1$ and $c \neq c'$.
(ii) Assume that there exists a non-trivial translation $t_1 : z \mapsto z + v_1$ and a rotation $\alpha : z \mapsto u(z-c)+c$ leaving the tiling invariant. Then, the composition

$$t_2 := \alpha \circ t_1 \circ \alpha^{-1} : z \mapsto z + uv_1 \tag{2.4}$$

is a symmetry as well. Call $v_2 := uv_1$. If $u \neq \pm 1$, clearly v_1 and v_2 are linearly independent (over $\mathbb{R}$). ■

Lemma 2.11 *A tiling $\mathcal{T}$ with (at least) one bounded prototile cannot have arbitrarily small translational symmetries. (That is: there exists $\delta > 0$ such that any translation of a non-zero vector v with $\|v\| < \delta$ is not a symmetry of $\mathcal{T}$.)*

Proof Let $T \in \mathcal{T}$ be a bounded tile. Since it has non-empty interior, T contains a disk $D_r(z_0)$ for some $r > 0$ and $z_0 \in \mathbb{C}$. Let $v \in \mathbb{C}$ with $\|v\| < \delta := 2r$ and let T' be the image of T under the translation $z \mapsto z + v$.

The situation is illustrated in the following picture:

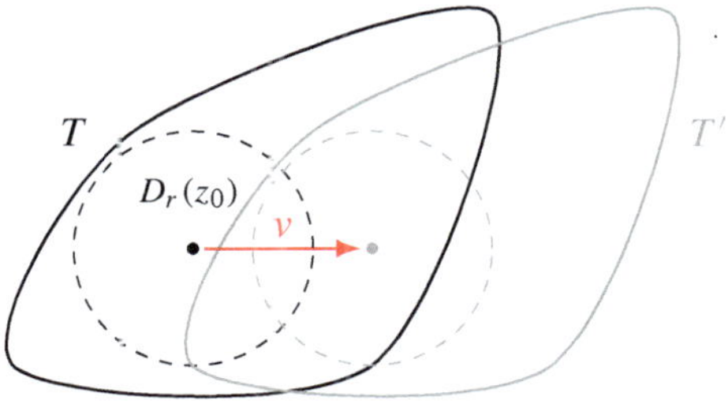

If the translation $z \mapsto z + v$ is a symmetry of the tiling, T' should be a tile in $\mathcal{T}$. But since $\|v\| < 2r$, one has $\mathring{T} \cap \mathring{T}' \neq \varnothing$ (since the disks overlap). This means that $T = T'$. But if $T = T + v$, then $T = T + nv$ for every $n \in \mathbb{Z}$, which means that either T is not bounded (contradicting the hypothesis) or $v = 0$. ■

Theorem 2.12 *Let $\mathcal{T}$ be a periodic tiling with (at least) one bounded prototile. If $\mathcal{T}$ has a rotational symmetry, then it must be n-fold with $n \in \{2, 3, 4, 6\}$.*

Proof Let $z \mapsto z + v_1$ and $z \mapsto u(z - c) + c$ be two non-trivial symmetries, with $u = e^{i\theta}$. By the argument in (2.4), the translations $z \mapsto z + v_2$ and $z \mapsto z + v_3$, with $v_2 = uv_1$ and $v_3 = uv_2$, are symmetries as well. Note that $v_1 + v_3 = 2\cos\theta \cdot v_2$.

Assume that $2\cos\theta$ is not an integer, call $\varepsilon \in]0, 1[$ its fractional part and k its integer part. Composing the translations in the direction of v_1, v_3 and $-kv_2$ we get a translational symmetry in direction of the vector $v_2' := v_1 + v_3 - kv_2 = \varepsilon v_2$. Conjugating it with a rotation by $\pm\theta$, we find that the tiling has also translational symmetries in the direction of the vectors $v_1' = \varepsilon v_1$ and $v_3' = \varepsilon v_3$. See Fig. 2.6.

We can repeat the argument to show, by induction, that the tiling has translational symmetries $z \mapsto z + \varepsilon^m v_i$, for $i \in \{1, 2, 3\}$ and all positive integers m. But $\varepsilon^m \to 0$ for $m \to +\infty$, and $\mathcal{T}$ cannot have arbitrarily small translational symmetries (Lemma 2.11). We got a contradiction.

Therefore, $2\cos\theta$ must be an integer. This means that $\cos\theta \in \{0, \pm\frac{1}{2}, -1\}$, and then $\theta = 2\pi/n$ with $n \in \{2, 3, 4, 6\}$. ■

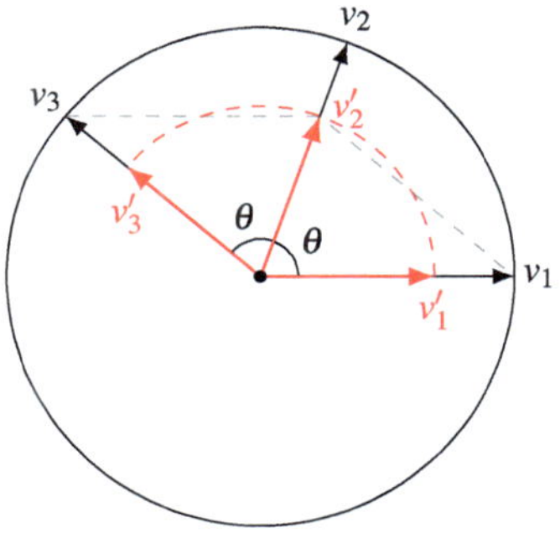

Fig. 2.6 Proof of Theorem 2.12

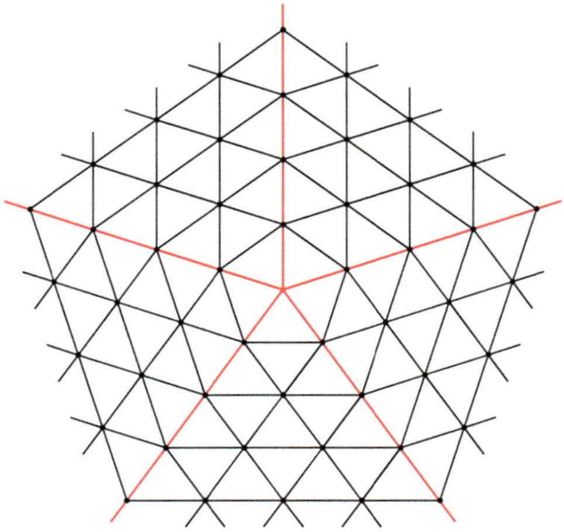

Fig. 2.7 A tiling with 5-fold symmetry

Non-periodic tilings and monohedral edge-to-edge tilings by non-regular polygons may have, of course, arbitrary rotational symmetries. For example, with the protoset:

$$\mathfrak{P} := \Bigg\{ \quad \Bigg\} \qquad (2.5)$$

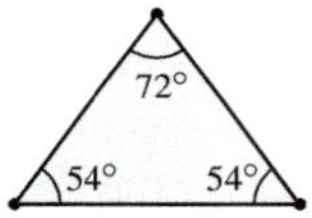

it is easy to produce a (non-periodic) tiling with 5-fold symmetry, cf. Fig. 2.7.

The tiling in Fig. 2.7 has 5-fold symmetry around a vertex, but the single tile has no non-trivial rotational symmetry. The following problem is still open:

Open problem 2.13 (See e.g. [DGS82]) *Under reasonable assumptions on the protoset (e.g. consisting of finitely many compact prototiles), find an example of tiling of the plane in which all the tiles have 5-fold symmetry (or n-fold with $n > 6$), or prove that such a tiling does not exist.*

Observe that the protoset (2.5) is not aperiodic. Indeed, with any triangle one can construct a parallelogram

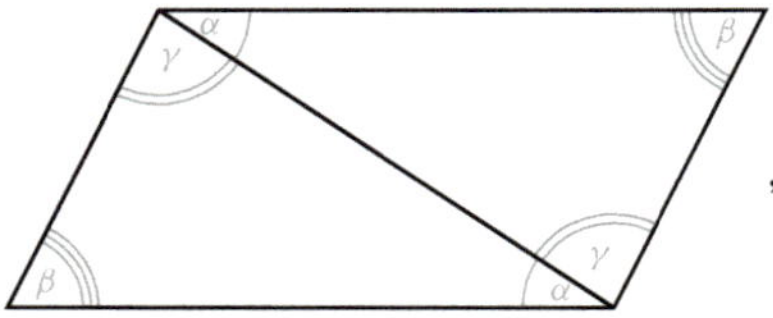

,

and then tile periodically the plane by placing a copy of this parallelogram on each cell of a suitable lattice.

2.4 Substitution Tilings

One way to construct bigger and bigger partial tilings is using a substitution rule. In this section we will illustrate this idea with two examples.

Let $\mathfrak{P}$ be a finite protoset and suppose there exists a scale factor $\lambda > 0$ such that, for every tile T with $[T] \in \mathfrak{P}$, the inflated tile λT is a finite union

$$\lambda T = \bigcup_\alpha T_\alpha,$$

where each T_α has class in $\mathfrak{P}$ and $\mathring{T}_\alpha \cap \mathring{T}_\beta = \varnothing$ for all $\alpha \neq \beta$.

In other words, for every T with $[T] \in \mathfrak{P}$ there exists a partial tiling $\mathcal{F}(T)$ with the same protoset $\mathfrak{P}$ covering exactly λT. It is enough to specify a tiling $\mathcal{F}(T)$ for each class $[T]$, and extend the map $\mathcal{F}$ to arbitrary tiles by declaring that, for all $(a, b) \in \mathbb{S}^1 \times \mathbb{C}$, if $T' = aT + b$ then $\mathcal{F}(T') = a\mathcal{F}(T) + \lambda b$. In this way we get a multivalued map $\mathcal{F}$ commuting with the action of rotations and rescaling translations by λ.

Given any partial tiling $\mathcal{T}$ covering exactly a set $X \subseteq \mathbb{C}$, the map $\mathcal{F}$ produces a partial tiling $\mathcal{F}(\mathcal{T})$ covering exactly the inflated set λX.

A simple example of this construction is given by the tiling (of the whole plane) in Fig. 2.7. Take $\lambda = 2$ and consider the substitution rule

where each triangle is similar to the one in (2.5), and the extra vertices are the midpoints of each segment.

If we start from a pentagon centered at the origin

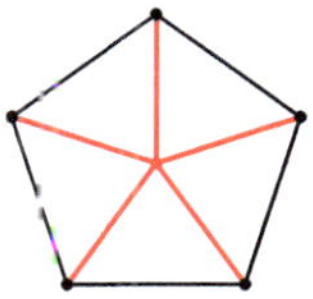

and apply, many times, a scaling by a factor 2 followed by the substitution rule above, we get bigger and bigger partial tilings with protoset (2.5). In fact, it is not difficult to verify that these partial tilings are each one contained into the next, and their union is the tiling of the plane in Fig. 2.7.

Substitution rules can be extremely complicated. The next example was introduced in [BFG07] (a long list can be found in [FHG]). Here the prototiles are a *domino* and a *kite*:

The domino is a rectangle of sides 1 and 2. The kite is obtained by gluing two right-angle triangles with sides 1 and 2, one mirror image of the other, along their diagonal. Assuming that three vertices of the kite have (real) coordinates given in the following picture

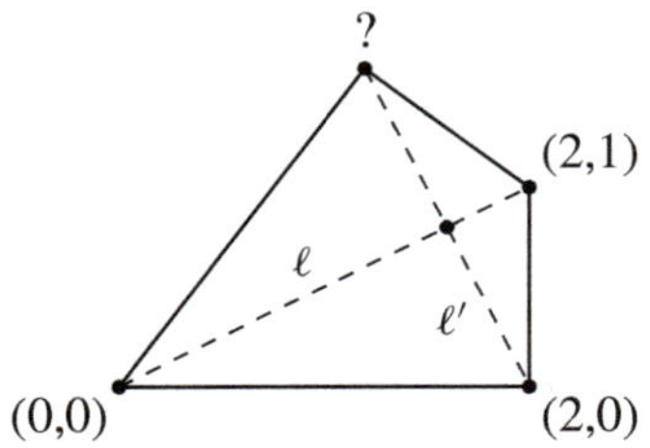

with a little Euclidean geometry one can find the coordinates of the fourth vertex.

Start with the Cartesian equation $x - 2y = 0$ of the line ℓ passing through the points $(0, 0)$ and $(2, 1)$ and the parametric equation

$$\{x_t := (t + 2, -2t) : t \in \mathbb{R}\}$$

of the orthogonal line ℓ' passing through the point $(2, 0)$. Using the formula for the point-line distance we compute the distance between x_t and ℓ, which is given by $|5t + 2|/\sqrt{5}$. The unknown point belongs to ℓ', and is at the same distance from ℓ than the point $(2, 0)$. This gives the condition $5t + 2 = -2$ for the unknown point, which then has coordinates:

$$x_{t=-4/5} = \frac{1}{5}(6, 8).$$

Using these coordinates one can check that the following substitution rule, for tiles inflated by a factor $\lambda = 5$, is well defined:

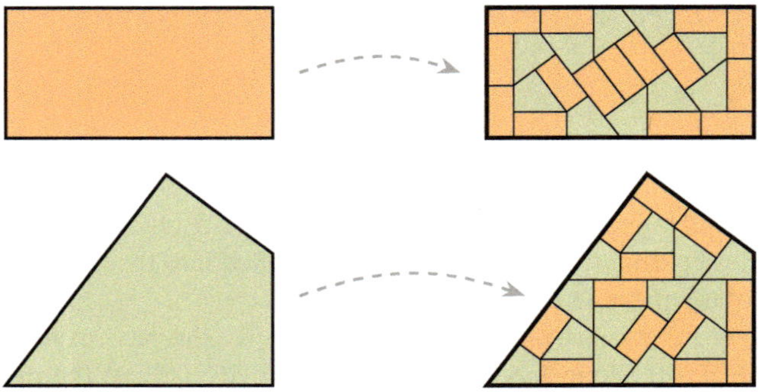

The example in Fig. 2.8 shows a kite and domino partial tiling obtained by applying the above rule twice to a domino tile. (Notice that the above rule produces tilings that are not edge-to-edge.)

The inflation and decomposition rule produces bigger and bigger partial tilings, which however are not contained one into the next. It is not obvious that "in the limit"

Fig. 2.8 A kite and domino tiling

(in a suitable sense) one gets a tiling of the plane. The existence of a tiling of the plane follows from a general theorem, called Extension Theorem, that is discussed in the next section. The next section is rather technical and the busy reader could skip it, as it is not strictly necessary in the study of Penrose tilings.

2.5 The Extension Theorem

Suppose we have a nested sequence of partial tilings

$$\mathcal{T}_0 \subseteq \mathcal{T}_1 \subseteq \mathcal{T}_2 \subseteq \dots \tag{2.6}$$

all with the same protoset $\mathfrak{P}$. Their union

$$\bigcup_{n\geq 0} \mathcal{T}_n$$

is then still a partial tiling with protoset $\mathfrak{P}$. If the partial tilings cover bigger and bigger concentric disks $D_{r_0}(z_0)$, $D_{r_1}(z_0)$, $D_{r_2}(z_0)$, ..., with $r_n \to +\infty$, then their union covers the whole plane:

$$\bigcup_{n\geq 0} \mathcal{T}_n \supseteq \bigcup_{n\geq 0} D_{r_n}(z_0) = \mathbb{C}.$$

However, if we have a sequence of partial tilings (with the same protoset $\mathfrak{P}$) not contained each one in the next, their union will not be a tiling and it is not obvious that a tiling of the plane (with protoset $\mathfrak{P}$) exists. The Extension Theorem guarantees, under reasonable hypothesis on the protoset, the existence of such a tiling.

Theorem 2.14 (Extension Theorem) *Let $\mathfrak{P}$ be a finite set of compact prototiles. If there exists a partial tiling with protoset $\mathfrak{P}$ covering $D_r(0)$ for all $r > 0$, then there exists a tiling of the plane with protoset $\mathfrak{P}$.*

The purpose of this section is to prove this theorem. A main ingredient is the topology on compact subsets of a metric space induced by the Hausdorff distance.

Note that if we can construct a partial tiling covering a disk $D_r(z_0)$, by translation we can construct one covering $D_r(0)$ as well. In Theorem 2.14 it is not important that the disks are centered at 0, or even concentric. In the case of a nested sequence (2.6), on the other hand, it is important that the disks are concentric. It is easy to produce nested sequences of disks (with $r_n \to +\infty$), whose union is not $\mathbb{C}$. For example, the union of all disks with integer radius tangent to the horizontal axis at the origin is the open half-plane:

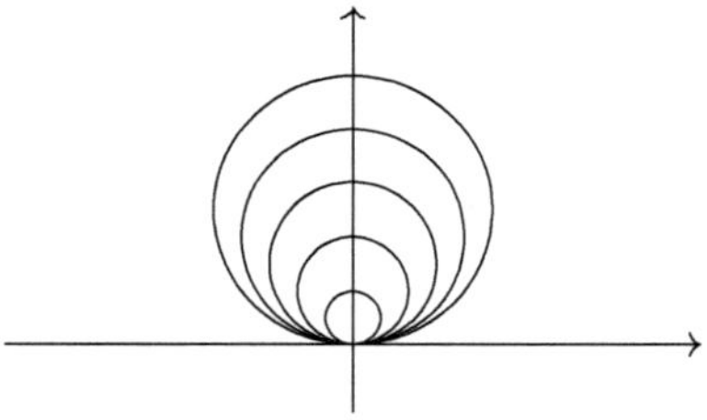

And if we translate each tiling to make the disks concentric, the translated tilings may not be included one into the next, and their union may not be a tiling.

In Theorem 2.14 both the hypothesis of finiteness of $\mathfrak{P}$ and compactness of the prototiles are necessary. It is easy to produce counterexamples where the theorem fails if we drop one of these two hypotheses.

Let us start with compactness. Suppose we have a single unbounded prototile, given by a closed strip which is infinite on one side and has a hole on the other side:

$\mathfrak{P} :=$ 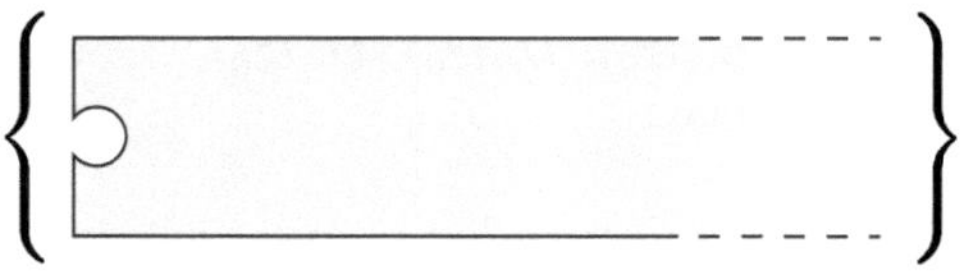 (2.7)

With this, we can construct partial tilings covering arbitrarily big disks:

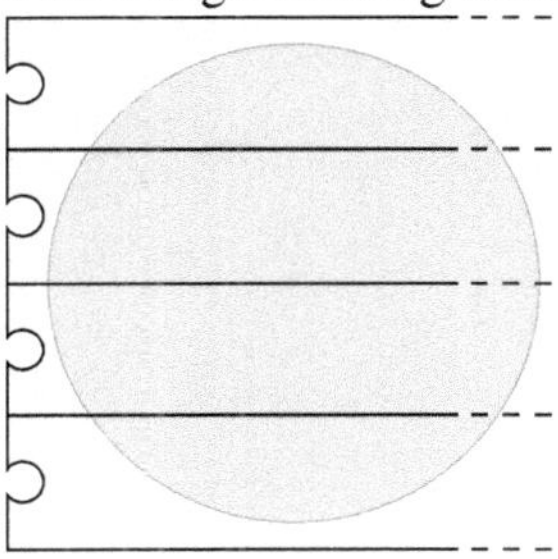

On the other hand, there is no tiling of the plane with protoset (2.7), as there is no way to fill the hole in a tile using another tile.

Next, consider the protoset with tiles constructed by carving a small hole (of fixed size) on squares with edges of integer length:

$\mathfrak{P}' :=$ 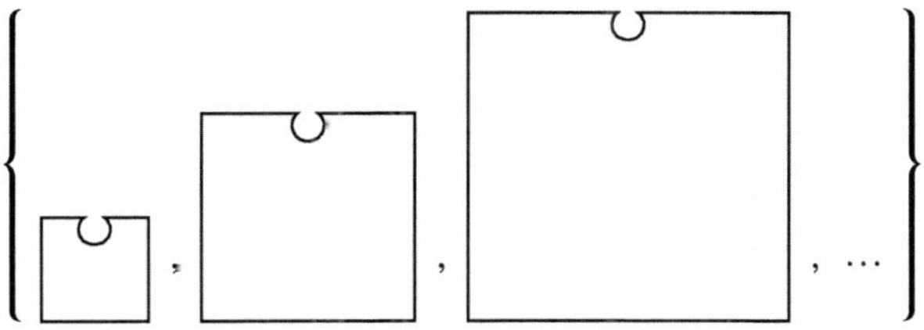 . (2.8)

By the same argument above, there is no tiling of the plane with protoset (2.8). But we can cover arbitrarily big disks using a single tile, as we have arbitrarily big tiles at our disposal.

2.5.1 The Hausdorff Distance

For a review of metric spaces and the Hausdorff distance one can see, e.g., [BBI01]. Here we recall only the basic definitions and properties.

Let X, Y be non-empty subsets of a metric space (M, d). For $x \in X$ we define the point-set distance as

$$d(x, Y) := \inf_{y \in Y} d(x, y) .$$

We define the eccentricity as

$$e_X(Y) := \sup_{x \in X} d(x, Y) .$$

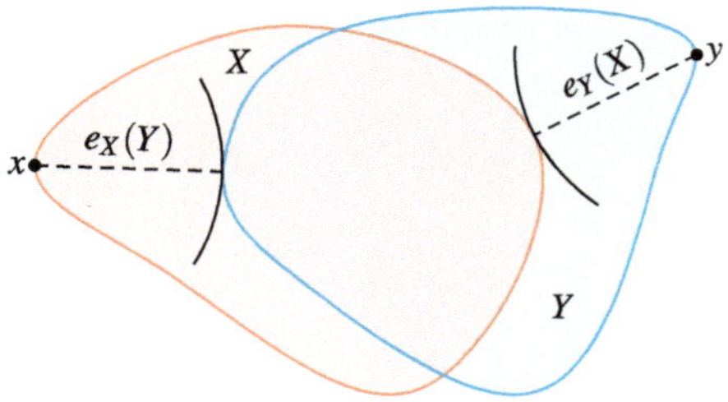

Fig. 2.9 Hausdorff distance between two subsets of the Euclidean plane

The Hausdorff distance is given by

$$d(X, Y) := \max \left\{ e_X(Y), e_Y(X) \right\} .$$

An illustration for subsets of the plane, with Euclidean distance, is in Fig. 2.9.

The Hausdorff distance is symmetric, reflexive ($d(X, X) = 0$ for all X) and satisfies the triangle inequality. However, it can be infinite and one can have $d(X, Y) = 0$ even if $X \neq Y$. The distance $d(X, Y)$ is finite if both X and Y are bounded, and it is zero if and only if X and Y have the same closure. As a consequence, the Hausdorff distance is a metric on the set $K(M)$ of all non-empty compact subsets of M.

We will always think of $K(M)$ as a topological space with topology induced by the Hausdorff distance (this is the *Vietoris topology* of $K(M)$, see [Bee93]).

Notice that, for every $x \in M$ and every non-empty $Y, Z \subseteq M$,

$$d(x, Z) \leq d(x, Y) + d(Y, Z). \tag{2.9}$$

Lemma 2.15 *Let $\delta > 0$, $p \in M$ and $(A_i)_{i \geq 0}$ a sequence of subsets that converges to A. If $d(p, A) \geq \delta$, then there is $i_0 \geq 0$ such that*

$$d(p, A_i) \geq \delta/2 \; \forall \, i \geq i_0. \tag{2.10}$$

Proof Let $\varepsilon > 0$. There exists $i_0 \geq 0$ such that $d(A, A_i) \leq \varepsilon$ for all $i \geq i_0$, which means that

$$\delta \leq d(p, A) \overset{(2.9)}{\leq} d(p, A_i) + d(A_i, A) \leq d(p, A_i) + \varepsilon.$$

For $\varepsilon = \delta/2$, we get (2.10). ∎

2.5.2 Group Actions

In the rest of this chapter, the metric space will be $M := \mathbb{C}$ equipped with norm-distance.

Lemma 2.16 *For every non-empty compact subset X of $\mathbb{C}$ and every $(a, b) \in \mathbb{C}^\times \times \mathbb{C}$ one has*

$$d(X, aX + b) \le r_X |a - 1| + |b|,$$

where $r_X := \sup_{x \in X} |x|$.

Proof Call $Y := aX + b$. For all $x \in X$,

$$d(x, Y) \le d(x, ax + b) = |ax + b - x| \le |a - 1| \cdot |x| + |b|.$$

Hence $e_X(Y) \le |a - 1| r_X + |b|$.

Similarly, for all $y \in Y$ and called $x' := a^{-1}(y - b)$, one has

$$d(X, y) \le d(x', y) = d(x', ax' + b) \le |a - 1| \cdot |x'| + |b|.$$

Therefore, $e_Y(X) \le |a - 1| r_X + |b|$ as well. ∎

Proposition 2.17 *For every non-empty compact subset $X \subset \mathbb{C}$, the function*

$$\mathbb{C}^\times \times \mathbb{C} \to K(\mathbb{C}), \qquad (a, b) \mapsto aX + b, \tag{2.11}$$

is continuous.

Proof Call $g := (a, b)$ and $f(g) := aX + b$. With obvious notations:

$$\begin{aligned} e_{f(g_1)}(f(g_2)) &= \sup_{x \in X} \inf_{y \in X} d(a_1 x + b_1, a_2 y + b_2) \\ &= |a_1| \sup_{x \in X} \inf_{y \in X} d(x, a_1^{-1}(a_2 y + b_2 - b_1)) = |a_1| e_X(Y), \end{aligned}$$

where $Y := aX + b$ with $a := a_1^{-1} a_2$ and $b := a_1^{-1}(b_2 - b_1)$. Similarly $e_{f(g_2)}(f(g_1)) = |a_1| e_Y(X)$, and then:

$$d\big(f(g_1), f(g_2)\big) = |a_1| d(X, Y).$$

Using Lemma 2.16:

$$d\big(f(g_1), f(g_2)\big) \le |a_1||a - 1| r_X + |a_1||b| = |a_1 - a_2| r_X + |b_1 - b_2| \le k_X\, d(g_1, g_2),$$

where $k_X := \sqrt{1 + r_X^2}$ and in the last step we used the Cauchy-Schwarz inequality. Observe that k_X is constant (X is fixed), hence f is Lipschitz continuous, which implies continuous. ■

2.5.3 The Extension Theorem

We now develop the tools that are needed to prove the Extension Theorem. The first one is the Selection Lemma, that in a nutshell states that: (i) given a sequence of tiles with finitely many shapes, we can pass to a subsequence where all the tiles have the same shape; (ii) if all these tiles have a point in common (see Fig. 2.10), then we can pass to a subsequence convergent to a tile with the same shape. The main ingredient in the proof is the continuity of the action of direct isometries with respect to the Vietoris topology.

Recall that two sets $X, Y \subseteq \mathbb{C}$ are called *equivalent* if they can be transformed one into the other with a direct isometry.

Lemma 2.18 (Selection Lemma) *Let $\mathfrak{F}$ be a finite collection of non-empty compact subsets of $\mathbb{C}$, $z_0 \in \mathbb{C}$ and $(S_i)_{i\geq 0}$ a sequence of subsets of $\mathbb{C}$ each equivalent to a set in $\mathfrak{F}$. If $z_0 \in S_i \ \forall\, i \geq 0$, then there is a convergent subsequence whose limit S is also congruent to a set in $\mathfrak{F}$.*

Proof Since $\mathfrak{F}$ is finite, for at least one $T \in \mathfrak{F}$ there must be infinitely many elements in the sequence that are equivalent to T. Pass to this subsequence, which is of the form

$$\left(a_i T + b_i\right)_{i\geq 0}$$

for some $g_i := (a_i, b_i) \in \mathbb{S}^1 \times \mathbb{C}$.

Since the map $\mathbb{S}^1 \times \mathbb{C} \to K(\mathbb{C})$ induced by (2.11) is continuous, it is enough to prove that the sequence $(g_i)_{i\geq 0}$ in $\mathbb{S}^1 \times \mathbb{C}$ has a convergent subsequence, and then call $S := aT + b$ with $g := (a, b)$ the limit of such a subsequence.

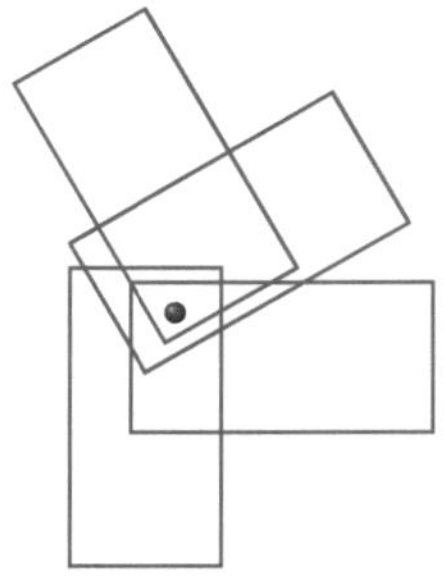

Fig. 2.10 The Selection Lemma

For all $i \geq 0$, one has $z_0 \in a_i T + b_i$, and then $a_i^{-1}(z_0 - b_i) \in T$. Let $k := \sup_{z \in T} |z|$ (this is finite, since T is compact, hence bounded). Since $|a_i| = 1$, it must be

$$|b_i - z_0| \leq k.$$

Thus, all elements $g_i := (a_i, b_i)$ of our sequence belong to a compact subset of $\mathbb{S}^1 \times \mathbb{C}$, the Cartesian product of a unit circle (since $|a_i| = 1$) and a closed disk of radius k centered at z_0. But in a metric space, compact is equivalent to sequentially compact. Hence the sequence $(g_i)_{i \geq 0}$ has a convergent subsequence. (Alternatively: by Bolzano-Weierstrass theorem every bounded sequence in $\mathbb{C}^2 \cong \mathbb{R}^4$ has a convergent subsequence.) ■

Lemma 2.19 *Let $\mathfrak{F}$ be a finite collection of non-empty compact subsets of $\mathbb{C}$, and $(A_i)_{i \geq 0}$ and $(B_i)_{i \geq 0}$ two sequences of subsets of $\mathbb{C}$ each congruent to a set in $\mathfrak{F}$ and convergent to sets A and B, respectively. If $\mathring{A} \cap \mathring{B} \neq \varnothing$, then there is a strictly increasing function $\mathbb{N} \to \mathbb{N}$, $n \mapsto i_n$ such that $\mathring{A}_{i_n} \cap \mathring{B}_{i_n} \neq \varnothing$ for all $n \geq 0$.*

Proof By the same argument used in the proof of the Selection Lemma, after passing to a subsequence we can assume that the sets are all of the form $A_{i_n} = a_n A + c_n$ and $B_{i_n} = b_n B + d_n$ for some $(a_n, c_n), (b_n, d_n) \in \mathbb{S}^1 \times \mathbb{C}$ with $(a_n, c_n) \to (1, 0)$ and $(b_n, d_n) \to (1, 0)$. Since $\mathring{A} \cap \mathring{B} \neq \varnothing$, we can find an open ball $D_r(z_0)$ contained in $\mathring{A} \cap \mathring{B}$. Observe that

$$D_r(z_0 + c_n) \subseteq \mathring{A}_{i_n} \quad \text{and} \quad D_r(z_0 + d_n) \subseteq \mathring{B}_{i_n} \tag{2.12}$$

for all n (we used the fact that $aD_r(z_0) + b = D_r(z_0 + b)$ for all $(a, b) \in \mathbb{S}^1 \times \mathbb{C}$). For all n greater than or equal to some n_0, the distance between the centers of the two disks in (2.12) satisfies

$$d(z_0 + c_n, z_0 + d_n) = |c_n - d_n| \leq |c_n| + |d_n| < r,$$

which means that the two disks overlap, and the corresponding sets $\mathring{A}_{i_n}$ and $\mathring{B}_{i_n}$ have non-empty intersection. After a reparametrization, we get $n_0 = 0$. ■

Lemma 2.20 *Let $\mathfrak{P}$ be a finite set of compact prototiles. Let $\mathcal{T} = \{T_\alpha\}$ be any partial tiling with prototiles in $\mathfrak{P}$. Then:*

(i) for every $r > 0$ and every $z_0 \in \mathbb{C}$, there are only finitely many tiles in $\mathcal{T}$ that have non-empty intersection with $D_r(z_0)$;
(ii) $\bigcup_\alpha T_\alpha$ is a closed subset of $\mathbb{C}$;
(iii) a point can be on the boundary of only finitely many tiles.

Proof (i) Every tile T contains an open disk of radius $r_{[T]}$ small enough (since open disks are a basis for the topology of $\mathbb{C}$, and the tiles have non-empty interior), and the radius clearly depends only on the congruence class $[T]$. Since $\mathfrak{P}$ is finite,

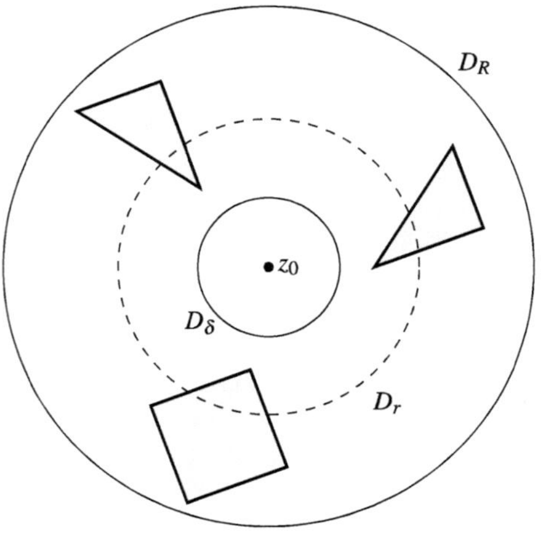

Fig. 2.11 Proof of Lemma 2.20

$\varepsilon := \min_{[T]\in\mathfrak{P}} r_{[T]} > 0$ exists. Choose $R > r$ big enough so that any tile that has non-empty intersection with $D_r(z_0)$ lies entirely inside $D_R(z_0)$ (it exists, since we have finitely many bounded prototiles). The set of distinct tiles that have non-empty intersection with $D_r(z_0)$ is finite: since each of them covers an area of at least $\pi\varepsilon^2$, they have non-overlapping interiors, and they are all contained in a disk of area πR^2, their number is at most $(R/\varepsilon)^2$. The situation is illustrated in Fig. 2.11.
(ii) Suppose $z_0 \notin \bigcup_\alpha T_\alpha$ and $r > 0$. The union of all the tiles intersecting $D_r(z_0)$ is closed (since they are finitely many closed sets), their complement is open and contains z_0. Therefore, there is a $\delta > 0$ small enough such that $D_\delta(z_0)$ has empty intersection with all the tiles intersecting $D_r(z_0)$. But if a tile doesn't intersect $D_r(z_0)$, cannot intersect its subset $D_\delta(z_0)$. Hence $D_\delta(z_0)$ is in the complement of $\bigcup_\alpha T_\alpha$. Every point in the complement (if any) is internal, hence the complement is open and $\bigcup_\alpha T_\alpha$ is closed.
(iii) Let $z_0 \in \mathbb{C}$. Any tile containing z_0 has non-empty intersection with $D_r(z_0)$, for any $r > 0$. The thesis then follows from point (i). ■

Lemma 2.20(i) implies that the collection $\mathcal{T}$ of tiles is locally finite (but in fact it is a much stronger property). To prove point (ii) we could have simply used the well known fact that the union of a locally finite collection of closed subsets of a topological space is itself closed.

We have now all the tools needed to prove the Extension Theorem.

Proof *of Theorem* 2.14. Every tile with class in $\mathfrak{P}$ contains an open disk of radius $\varepsilon_{[T]}$ small enough (since it has non-empty interior and open disks form a basis of the topology of $\mathbb{C}$). Since $\mathfrak{P}$ is finite, $\varepsilon := \sqrt{2}\min_{[T]\in\mathfrak{P}} \varepsilon_{[T]} > 0$ exists. Consider the lattice $\Lambda := \varepsilon\,\mathbb{Z}^2 \subset \mathbb{C}$ of complex numbers whose real and imaginary part are integer multiples of ε. Every tile with class in $\mathfrak{P}$, whatever its position on the plane is, must contain (at least) one point of the lattice Λ in its interior.

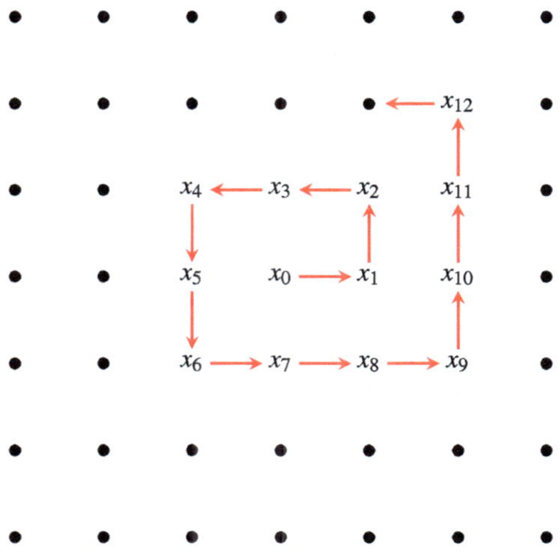

Fig. 2.12 Proof of Theorem 2.14

Number the points in Λ by spiraling outwards starting from $x_0 := 0$, like in Fig. 2.12. By assumption, for every $r > 0$ there exists a partial tiling $\mathcal{T}_r$ covering $D_r(0)$. For every $r > m\sqrt{2}$, $m \in \mathbb{N}$, the disk $D_r(0)$ contains the closed square with vertices $(\pm m, \pm m)$, and then the points $x_0, x_1, \ldots, x_{4m(m+1)}$ of the lattice Λ. Let m_r be the biggest integer less than $r/\sqrt{2}$. For each $r > 0$ and each $i \in \{0, 1, \ldots, m_r\}$, choose one tile in $\mathcal{T}_r$ containing the point x_i and call it $T_{r,i}$ (it exists by the argument above, but there may be more than one if the point is on the boundary of a tile). Observe that $\mathring{T}_{r,i} \cap \mathring{T}_{r,j} = \varnothing$ unless $T_{r,i} = T_{r,j}$, which may happen for some $i \neq j$ since the list $(T_{r,0}, T_{r,1}, \ldots, T_{r,m_r})$ may contain repetitions. Observe that, for every $i \in \mathbb{N}$, $T_{r,i}$ is well-defined if r is big enough.

Use the Selection Lemma to find an increasing divergent sequence $(r_n^0)_{n\geq 0}$ of radii such that $T_{r_n^0,0} \to T_0'$, where $[T_0'] \in \mathfrak{P}$. Use the Selection Lemma again to find a subsequence $(r_n^1)_{n\geq 0}$ such that $T_{r_n^1,1} \to T_1'$, where $[T_1'] \in \mathfrak{P}$, and observe that $T_{r_n^1,0} \to T_0'$ where T_0' is the same as before. Observe that if $\mathring{T}_0' \cap \mathring{T}_1' \neq \varnothing$, from Lemma 2.19 we deduce that $\mathring{T}_{r_n^1,0} \cap \mathring{T}_{r_n^1,1} \neq \varnothing$ for all n greater than some n_0 (possibly after passing to a subsequence). But this means $T_{r_n^1,0} = T_{r_n^1,1}$ for all $n \geq n_0$ (since $T_{r_n^1,0}$ and $T_{r_n^1,1}$ belong to a tiling), which implies $T_0' = T_1'$. Thus, either $T_0' = T_1'$ or $\mathring{T}_0' \cap \mathring{T}_1' = \varnothing$.

Repeating the argument, by induction one finds sequences of radii $(r_n^k)_{n\geq 0}$—one for every $k \geq 0$ and each one a subsequence of the previous one—and a sequence

$$(T_0', T_1', T_2', \ldots) \tag{2.13}$$

of tiles with class in $\mathfrak{P}$, with non-overlapping interiors, and such that

$$\begin{aligned} x_i &\in T_i' && \forall\, i \geq 0, \\ T_i' &= \lim_{n\to\infty} T_{r_n^k,i} && \forall\, 0 \leq i \leq k. \end{aligned}$$

It remains to show that the tiles (2.13) cover the plane. By contradiction, suppose there exists $z_0 \in \mathbb{C}$ which does not belong to any tile in (2.13). From Lemma 2.20(ii) we know that there exists an open disk $D_\delta(z_0)$ with empty intersection with all the tiles in (2.13), which means that z_0 is at distance at least δ from any tile T_i'. Thus, for every $0 \leq i \leq k$ the tile $T_{r_n^k,i}$ is at distance at least $\delta/2$ from z_0 if n is big enough (cf. Lemma 2.15).

Now, choose R big enough so that $z_0 \in D_R(0)$ and any tile (with class in $\mathfrak{P}$) containing z_0 is entirely contained in $D_R(0)$. Choose

$$k_0 \geq \max\left\{i : x_i \in D_R(0)\right\} \tag{2.14}$$

and choose n_0 big enough so that $r_{n_0}^{k_0} > R$. By the above choices of R and n_0, there exists a tile $S \in \mathcal{T}_{r_n^{k_0}}$ that contains z_0 for every $n \geq n_0$. Such a tile must contain a lattice point x_i in its interior, thus $S = T_{r_n^{k_0},i}$ for some i; since $x_i \in S \subseteq D_R(0)$, it follows from (2.14) that $i \leq k_0$.

Thus, for every $n \geq n_0$ there exists $0 \leq i \leq k_0$ such that $z_0 \in T_{r_n^{k_0},i}$, contradicting the fact that these tiles should be at distance $\geq \delta/2$ from z_0 if n is big enough. ■

Chapter 3
Robinson Triangles

Robinson triangles were introduced in [Rob75] to study Penrose tilings. In this book, we follow the logical rather than chronological order and start with triangles. To construct them, we cut a regular pentagon with unit sides in three parts:

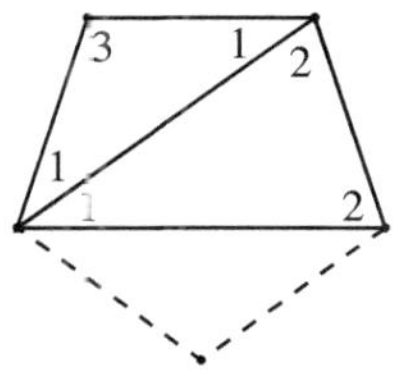

In this way, we get two isosceles triangles

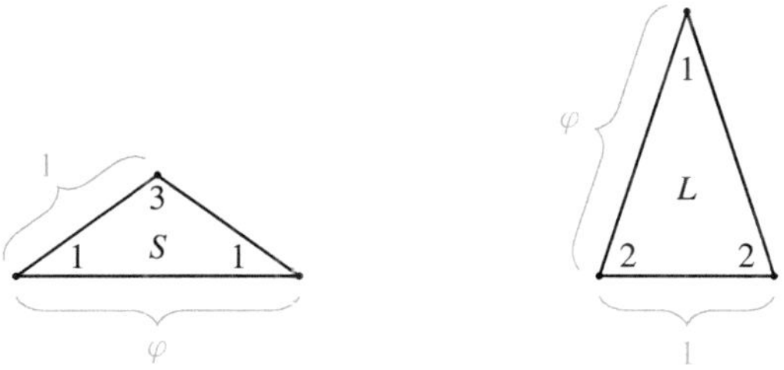

In the picture, the integers indicate the internal angles in multiples of $\pi/5$. The short edges have unit length, and the long ones have length given by the golden ratio $\varphi = \frac{1+\sqrt{5}}{2}$, whose properties are recalled in Appendix A.1.

We denote by S and L the congruence classes of these two triangles and call them "small" and "large", respectively.

The small triangle can be inscribed in the large one, and the large triangle can be inscribed in a small triangle rescaled by a factor φ, cf. (A.8):

F. D'Andrea, *A Guide to Penrose Tilings*,
https://doi.org/10.1007/978-3-031-28428-1_3

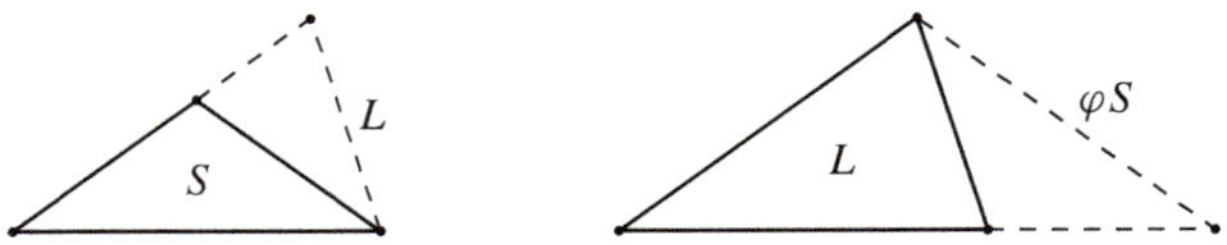

A triangle similar to L will be called a *golden triangle*, while a triangle similar to S will be called a *golden gnomon*.

With these two congruence classes of triangles, L and S, we now construct a set of four prototiles by adding a matching rule. We decorate each edge with an arrow:

$\mathfrak{P} :=$ 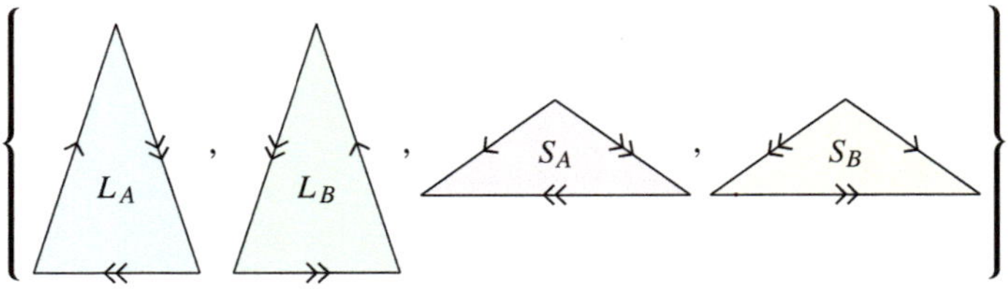 (3.1)

We colored the triangles for visual reasons only. We repeat that, in each triangle, all short edges have length 1 and all long edges have length φ. The only allowed partial tilings are edge-to-edge, and adjacent tiles can only share an edge with the same type of arrow and pointing in the same direction.

Observe that the B-prototiles are obtained from the A-prototiles by reflection. Some authors define a prototile as a congruence class of tiles: with that convention the set (3.1) would consist of two prototiles rather than four. However, when one enforces the matching rules by shape, cf. (3.5), it is no longer true that the set of prototiles is invariant under reflections. It is then better to keep the A and B prototiles distinct, and follow the convention that a prototile is an equivalence class of tiles under the action of direct isometries only.

We will refer to the congruence class of a prototile in (3.1), either L or S, as its *shape*. Because of the decomposition (1.1), prototiles with shape L are also called *semikites*, while those with shape S are called *semidarts* (cf. Chap. 4). For a given shape, the directions of the arrows determine what we call the *type* of the prototile, either A or B. Observe that from the shape and type one can reconstruct the markings, so that they could be omitted in the pictures (and often we will do it).

Definition 3.1 A tiling of the plane with protoset (3.1) satisfying the matching rule described above will be called a *Robinson tiling*.

At this stage, it is not obvious that Robinson tilings exist. If one starts constructing a tiling by hand attaching triangles, he will sooner or later reach a point where the partial tiling can no longer be extended (we will see many examples of illegal partial tilings). The Extension Theorem allows us to prove the existence of Robinson tilings without actually constructing them, cf. Sect. 3.1. In the next section we will see a constructive proof as well.

Observe that there is no way to glue together two copies of the same prototile, or two tiles with different shape but same type, respecting the matching rule. Two tiles with the same shape but different type can be joined in three different ways:

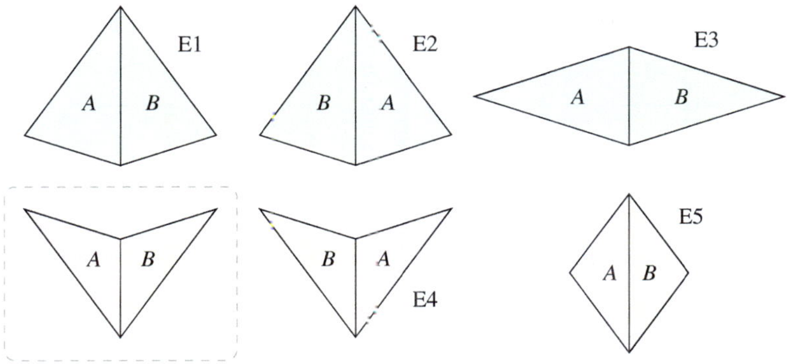

(3.2a)

Two tiles that differ both in shape and type can be glued in two ways, along either the short or the long edge with double arrow:

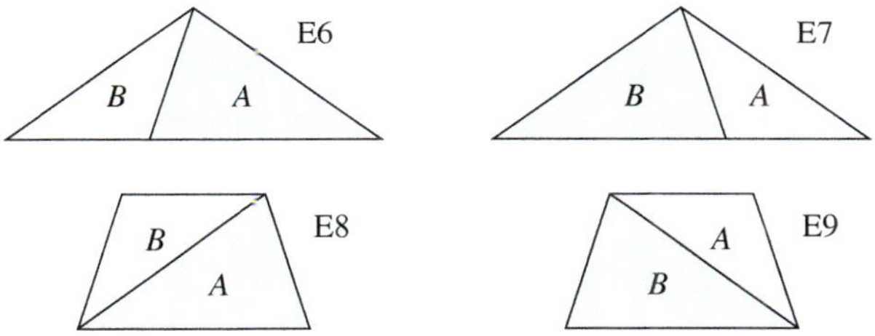

(3.2b)

We will see that the framed configuration in (3.2a) never appears in a tiling of the plane (cf. Lemma 3.7). The legal edge neighborhoods are labeled from E1 to E9 for future reference. In Appendix A.3 there is a page that one can copy and cut to experiment with Robinson's triangles.

3.1 Decomposition and Existence of Tilings

Let us consider the collection $\mathfrak{R}$ of all tiles (decorated with arrows) that are similar to a Robinson triangle (when we place triangles in the plane we now allow not only rotations and translations, but scalings as well). One can easily find a smooth parametrization of $\mathfrak{R}$: for each prototile in (3.1) choose a representative (that is: place the prototile in some position in the plane); every element of $\mathfrak{R}$ can now be specified by its shape and type, and the affine transformation used to place it in the plane. Thus:

$$\mathfrak{R} \cong \{L_A, L_B, S_A, S_B\} \times \mathrm{Aff}(\mathbb{C}).$$

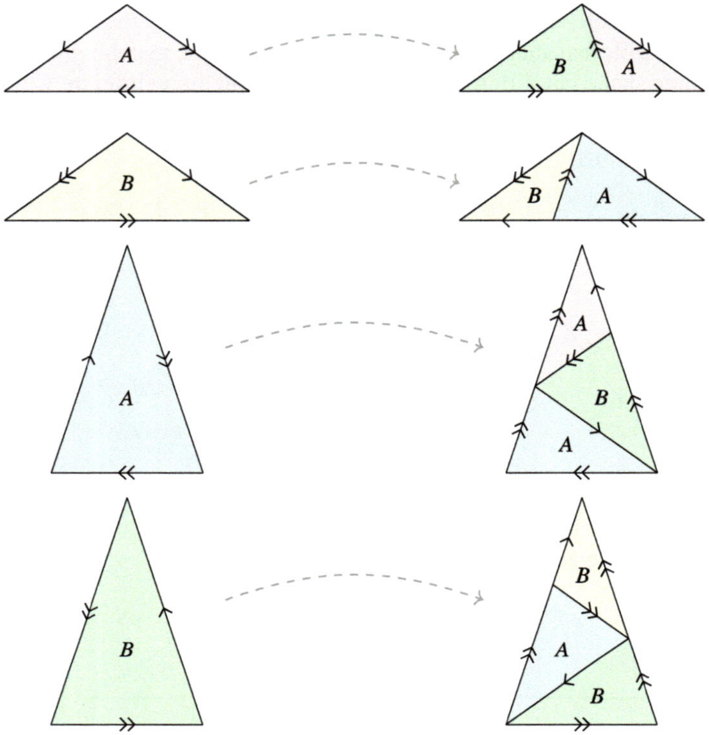

Fig. 3.1 Decomposition

We now define a multivalued function

$$\mathcal{F} : \mathfrak{R} \to \mathfrak{R}, \tag{3.3}$$

as follows. Given $T \in \mathfrak{R}$, first we apply a scaling of a factor φ. Then, depending on the triangle, we apply to φT the subdivision rule in Fig. 3.1 (without changing the position of the outer triangle).

The subdivision rule in Fig. 3.1 defines a unary operation on the set of all partial tilings with prototiles similar to Robinson triangles. We call this operation

Decomposition

In the whole book, when we write Decomposition with a capital letter we will always refer to this operation.

A careful inspection of the figures and a little Euclidean geometry show that, if T is a Robinson triangle rescaled by a factor λ, then $\mathcal{F}(T)$ is a collection of Robinson triangles rescaled by the same factor λ (the subdivision compensates the initial scaling by φ). The function $\mathcal{F}$ is multivalued, since it maps every S triangle to a pair of triangles, and every L triangle to a triple of triangles. It is clear from its definition that:

Lemma 3.2 *Decomposition commutes with similarity transformations.*

The map $\mathcal{F}$ allows to produce arbitrary big partial tilings by Robinson triangles, and prove the existence of tilings of the plane. The crucial step in the proof is the next lemma.

Lemma 3.3 *If $\mathcal{T}$ is a partial tiling by Robinson triangles covering exactly a set X, then $\mathcal{F}(\mathcal{T})$ is a partial tiling by Robinson triangles covering exactly the set φX.*

Proof The tricky part is to show that the subdivision process in Fig. 3.1 is compatible with the matching rule. One must check case by case, cf. (3.2a)–(3.2b), that if we look at two adjacent tiles and apply our map $\mathcal{F}$, the subdivision process gives a partial tiling still respecting the matching rule. This is obvious when the two tiles share a short edge, since short edges are not modified by $\mathcal{F}$. For long edges, the substitution rule (regardless of the tile we are considering) is:

and

For example:

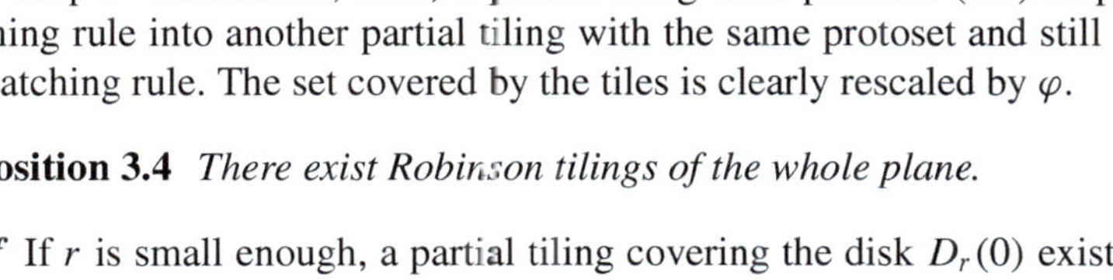

(3.4)

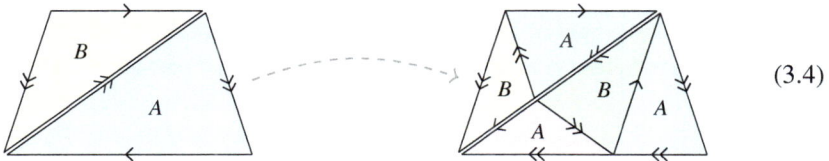

The map $\mathcal{F}$ transforms, then, a partial tiling with protoset (3.1) respecting the matching rule into another partial tiling with the same protoset and still respecting the matching rule. The set covered by the tiles is clearly rescaled by φ. ■

Proposition 3.4 *There exist Robinson tilings of the whole plane.*

Proof If r is small enough, a partial tiling covering the disk $D_r(0)$ exists: take for example $r = 1/4$ and cover the disk with a single tile. By Lemma 3.3, if we apply n times the map $\mathcal{F}$ to this partial tiling, we get a partial tiling covering $D_{\varphi^n r}(0)$. Since $\varphi > 1$, one has $\varphi^n r \to +\infty$ for $n \to +\infty$ and the Extension Theorem ensures the existence of a tiling of the plane with prototiles (3.1).

It remains to show that we get a tiling respecting the matching rule. Consider the protoset given by the following four prototiles, obtained from Robinson's triangles by slightly modifying the edges:

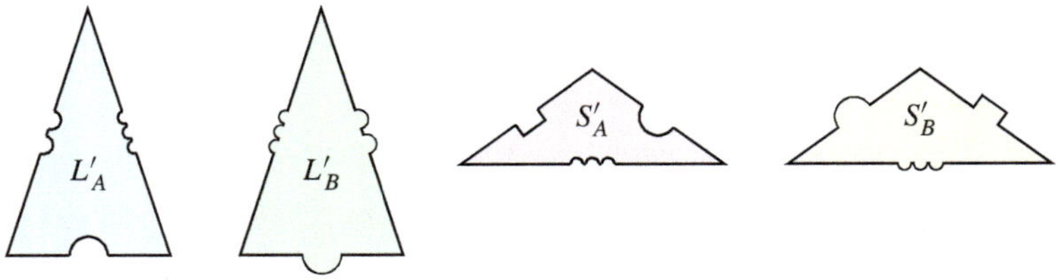

(3.5)

The edges are designed to enforce the matching rule. Any partial tiling by Robinson triangles covering $D_{\varphi^n r}(0)$ gives rise to a partial tiling with protoset (3.1) covering the same disk (maybe slightly smaller, so that the holes in the external edges don't cross the boundary of the disk). The Extension Theorem guarantees the existence of a tiling of the plane with protoset (3.5). In this tiling of the plane we replace each tile by the corresponding Robinson triangle and obtain a tiling $\mathcal{T}$ of the plane with prototiles $\{L_A, L_B, S_A, S_B\}$. Since the squiggly edges of the prototiles (3.5) were designed to enforce the matching rule, in the tiling $\mathcal{T}$ the matching rule for Robinson's triangles is respected. ■

In the next subsections we will give an alternative proof of Proposition 3.4 by constructing explicitly several Robinson tilings.

3.1.1 The Cartwheel

Figure 3.2 shows the first iterations of the inflation and subdivision algorithm when the starting point is the edge neighborhood E1, that we place in the plane pointing down and with the point of intersection of the angle bisectors at the origin (the point marked with a red cross in the picture). We call $\mathcal{C}_0$ this first partial tiling, and define inductively

$$\mathcal{C}_n := \mathcal{F}(\mathcal{C}_{n-1}), \qquad \forall\, n \geq 1.$$

A partial tiling congruent to $\mathcal{C}_1$ is called an ACE, while a partial tiling congruent to $\mathcal{C}_n$ will be called a *Cartwheel of order n*. Observe that the ace is not congruent to the big kite in Proposition 3.9 (the ace is not a vertex neighborhood).

We see from the picture that

$$\mathcal{C}_0 \subset \mathcal{C}_2. \tag{3.6}$$

Notice that $\mathcal{C}_0$ is not just congruent to a subset of $\mathcal{C}_2$, but really contained in $\mathcal{C}_2$. To convince ourselves that the small kite in $\mathcal{C}_2$ is in the correct position in the plane, we can observe that the set $\cup\mathcal{C}_2$ and the small kite E1 inside it have the same point of intersection of the angle bisectors, corresponding to the origin of $\mathbb{R}^2$.

The application of $\mathcal{F}^n$ to both sides of (3.6) yields

$$\mathcal{C}_n \subset \mathcal{C}_{n+2}, \quad \forall\, n \in \mathbb{N}.$$

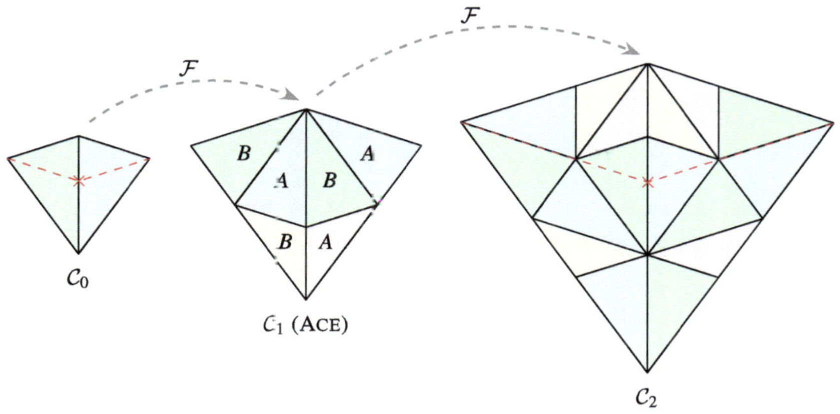

Fig. 3.2 The Cartwheel

Since we have a nested sequence of partial tilings, their union

$$\mathcal{C} := \bigcup_{n \in 2\mathbb{N}+1} \mathcal{C}_n$$

is a partial tiling as well. It covers the plane since the partial tilings $\mathcal{C}_n$ cover arbitrarily big disks centered at the origin. The tiling $\mathcal{C}$ is called *Cartwheel* (in fact, sometimes we will use the term "Cartwheel" to refer to any tiling congruent to $\mathcal{C}$: it will be clear from the context if by Cartwheel we mean $\mathcal{C}$ or a tiling congruent to it).

The Cartwheel is invariant under reflection across the vertical axis. Moreover, since $\mathcal{F}^2(\mathcal{C}_n) = \mathcal{C}_{n+2} \supset \mathcal{C}_n$, the tiling $\mathcal{F}^2(\mathcal{C}) = \bigcup_{n \in 2\mathbb{N}+3} \mathcal{C}_n$ both contains and is contained in $\mathcal{C}$. Thus:

$$\mathcal{F}^2(\mathcal{C}) = \mathcal{C}.$$

3.1.2 The Sun and the Star

We now apply the algorithm (3.3) to a partial tiling given by ten L triangles incident in the origin, i.e. the "wheel" $\mathcal{T}_0$ in Fig. 3.3. If we apply $\mathcal{F}$ we obtain the partial tiling $\mathcal{T}_1$ in Fig. 3.3 (with a "star" inside). Next, we apply $\mathcal{F}$ and then a counterclockwise rotation of $\pi/5$ around the origin, and obtain the tiling $\mathcal{T}_2$ in Fig. 3.3. Thus, $\mathcal{T}_2 = e^{\pi i/5}\mathcal{F}^2(\mathcal{T}_0)$. We continue with inflation, subdivision, and a rotation every other step, and get a sequence of partial tilings defined inductively by:

$$\mathcal{T}_{n+2} = e^{\pi i/5}\mathcal{F}^2(\mathcal{T}_n), \qquad \forall\, n \in \mathbb{N}.$$

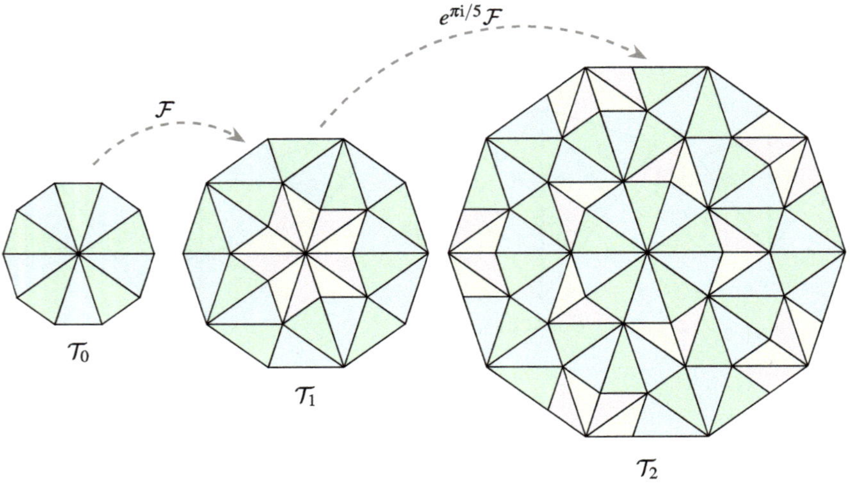

Fig. 3.3 Tilings with 5-fold symmetry

We see from the picture that $\mathcal{T}_0 \subset \mathcal{T}_2$. If we apply $(e^{\pi i/5}\mathcal{F}^2)^n$ to both sides of this inclusion, we get

$$\mathcal{T}_n \subset \mathcal{T}_{n+2} \tag{3.7}$$

for all even n. We have a nested sequence of partial tilings covering bigger and bigger concentric disks: their union is then a tiling of the plane.

Applying $\mathcal{F}$ to both sides of the inclusion $\mathcal{T}_0 \subset \mathcal{T}_2$ one also finds out that $\mathcal{T}_1 \subset \mathcal{T}_3$, so that (3.7) holds for odd n too. We get a second nested sequence of partial tilings, whose union tiles the plane.

We call $\mathcal{T}_{\text{even}} := \bigcup_{n\in\mathbb{N}} \mathcal{T}_{2n}$ the *Sun* and $\mathcal{T}_{\text{odd}} := \bigcup_{n\in\mathbb{N}} \mathcal{T}_{2n+1}$ the *Star* tiling.

Proposition 3.5 *The Sun and the Star tilings have 5-fold symmetry.*

Proof $\mathcal{T}_0$ and $\mathcal{T}_1$ have 5-fold symmetry (see Fig. 3.3):

$$\mathcal{T}_0 = e^{2\pi i/5}\mathcal{T}_0 \qquad \text{and} \qquad \mathcal{T}_1 = e^{2\pi i/5}\mathcal{T}_1.$$

If we apply $(e^{\pi i/5}\mathcal{F}^2)^n$ both sides of the above equalities, since $\mathcal{F}$ commutes with rotations, we find that $\mathcal{T}_n = e^{2\pi i/5}\mathcal{T}_n$ for all $n \in \mathbb{N}$. ■

It follows from Theorem 2.12 that the Sun and the Star tilings are non-periodic. In the next section we will prove that all Robinson tilings are non-periodic, i.e. (3.1) is an aperiodic protoset.

Lemma 3.6 *The Sun and the Star tilings are not equivalent.*

Proof The center of symmetry in a non-periodic tiling is unique (Lemma 2.10(i)), and any isometry transforming the Sun into the Star should map the center of symmetry of the first to the center of symmetry of the second as well. Thus, it should transform a wheel into a star:

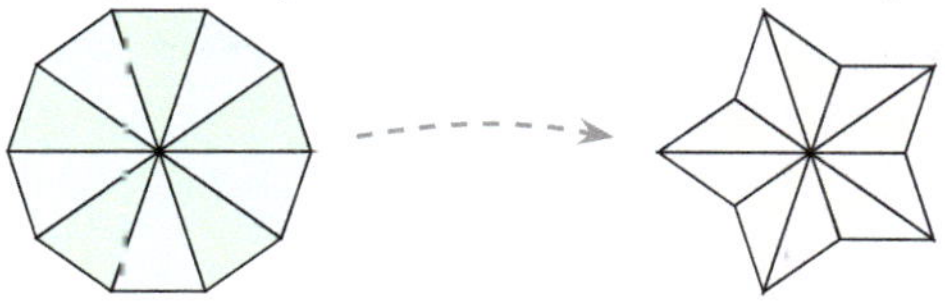

which is clearly impossible. ■

Similarly to the Cartwheel, one proves that $e^{\pi i/5}\mathcal{F}^2(\mathcal{T}_{\text{even}}) = \mathcal{T}_{\text{even}}$, which combined with the 5-fold symmetry gives

$$\mathcal{F}^4(\mathcal{T}_{\text{even}}) = \mathcal{T}_{\text{even}}.$$

Finally, we may observe that $\mathcal{T}_{\text{odd}} = \mathcal{F}(\mathcal{T}_{\text{even}})$.

3.1.3 The Golden Triangle and the Golden Gnomon

Figure 3.4 shows the first four iterations of the inflation and subdivision algorithm when the starting point is an L_A triangle. Call $\mathcal{T}_0$ the initial partial tiling and

$$\mathcal{T}_n := \mathcal{F}^{4n}(\mathcal{T}_0)\ , \quad n \geq 1.$$

As we can see in Fig. 3.4, $\mathcal{T}_1$ contains a copy of $\mathcal{T}_0$. We can place the initial tile in the plane in such a way that one has a set inclusion. We let the origin coincide with the intersection of the half-lines (red in the picture) from the basis vertices of L_A and with slope 54° and 126°. These lines cut the basis angles in two parts, one triple of the other. With a little Euclidean geometry one proves that the corresponding inflated lines in $\mathcal{T}_1$ cross the basis vertices of an L_A tile, proving that the position of this tile is the same as in $\mathcal{T}_0$. Thus $\mathcal{T}_0 \subset \mathcal{T}_1$ and applying $\mathcal{F}^{4n}$ to both sides of this inclusion we get

$$\mathcal{T}_n \subset \mathcal{T}_{n+1}, \quad \forall\, n \geq 0.$$

We have a nested sequence of partial tilings covering bigger and bigger disks centered at the origin. Their union $\mathcal{T} := \bigcup_{n\geq 1} \mathcal{T}_n$ is then a tiling of the plane that we call, for future reference, the *Golden Triangle* (like the tile). Similarly to the Cartwheel, Sun and Star tilings one proves that:

$$\mathcal{F}^4(\mathcal{T}) = \mathcal{T}.$$

There is one more explicit construction that is worth mentioning, very similar to the previous one. The only delicate point is where to place the origin in order to get a nested sequence of partial tilings. Figure 3.5 shows the first few iterations of the inflation and subdivision algorithm when the starting point is an S_A triangle.

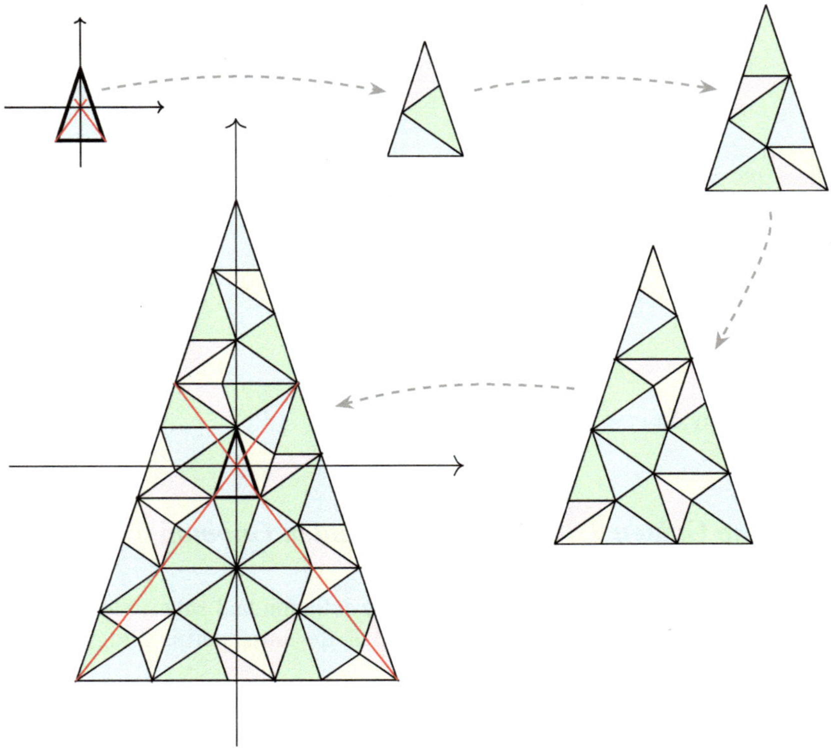

Fig. 3.4 Iterations from a single L triangle

Call $\mathcal{G}_0$ the first partial tiling in the picture, obtained by applying $\mathcal{F}$ to an S_A tile that we place with its barycenter at the origin. For $n \geq 1$, let $\mathcal{G}_n := \mathcal{F}^{4n}(\mathcal{G}_0)$. In $\mathcal{G}_1$ there is a framed S triangle that has the same barycenter as the huge outer triangle (the origin of $\mathbb{R}^2$), proving that we have a set inclusion $\mathcal{G}_0 \subset \mathcal{G}_1$. The rest of the argument is the same as for the Golden Triangle: $\mathcal{G} := \bigcup_{n \geq 1} \mathcal{G}_n$ is a tiling of the plane that we call, for future reference, the *Golden Gnomon* (like the tile). This tiling satisfies

$$\mathcal{F}^4(\mathcal{G}) = \mathcal{G}$$

like the Golden Triangle.

The Golden Gnomon is a good example to show how a different position of the initial partial tiling in the plane can produce different (inequivalent) tilings. Repeat the same construction above, starting with $\mathcal{G}'_0$ given by an S_A tile, but now place the tile so that the origin coincides with the blue point in the last picture in Fig. 3.5, which now represents $\mathcal{G}'_1 := \mathcal{F}^5(\mathcal{G}'_0)$. Again $\mathcal{G}'_0 \subset \mathcal{G}'_1$ and we get another tiling $\mathcal{G}'$ of the whole plane, this time satisfying

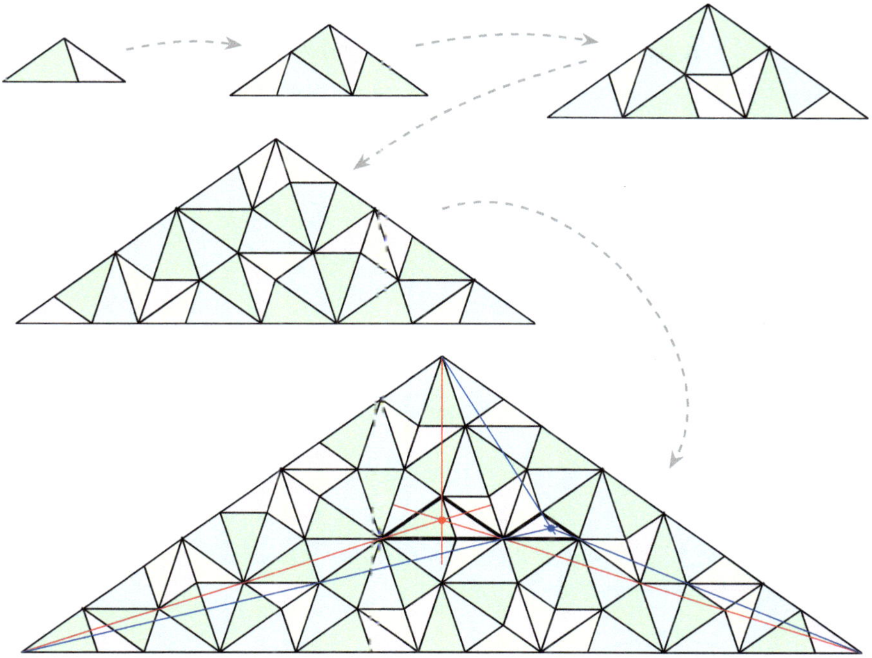

Fig. 3.5 Iterations from a single S triangle

$$\mathcal{F}^5(\mathcal{G}') = \mathcal{G}'.$$

We will see in Sect. 3.4 that $\mathcal{G}'$ is not equivalent to $\mathcal{G}$.

3.2 Vertex Neighborhoods

The aim of this section is to classify all possible vertex neighborhoods in a Robinson tiling. We need some preliminary observations.

Lemma 3.7 *In a Robinson tiling, two tiles with class S_A and S_B never share a short edge with a double arrow.*

Proof By contradiction, assume that there is a tile with class S_A attached to a tile with class S_B along the short edge with a double arrow. In a tiling of the whole plane, every edge of a tile is shared with another tile. The same tile S_A should have a tile attached to the edge with a single arrow, and the only possibility is that it is a tile S_B, like in the picture:

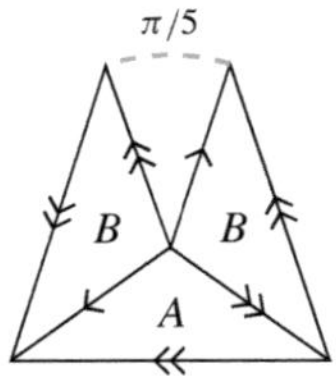

Now, the only way to fill the white space is with the sharp edge of an S or L triangle, but it is impossible to do it respecting the matching rule. ■

Remark 3.8 In a Robinson tiling:

1. an S triangle has always an L triangle attached along the short edge with a double arrow,
2. an S triangle has always another S triangle attached along the edge with a single arrow,
3. an L triangle has always another L triangle attached along the edge with a single arrow.

Proposition 3.9 *In a Robinson tiling (up to a direct isometry) only the following seven vertex neighborhoods are possible:*

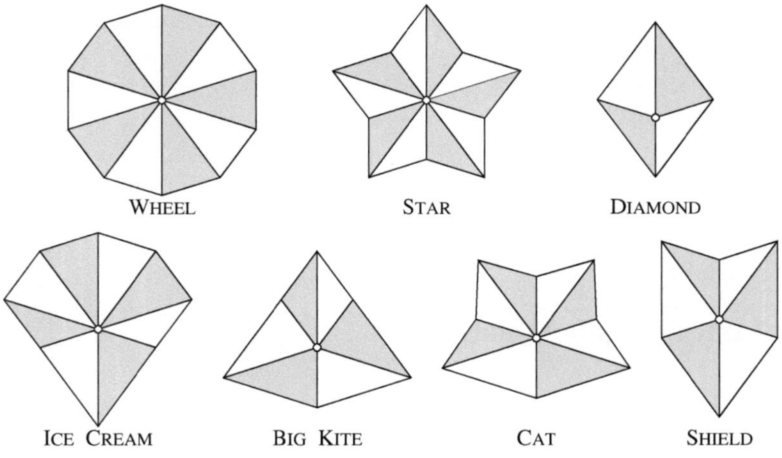

(Here the tiles of type A are in light gray, and those of type B are in dark gray.)

Proof For each of the nine edge neighborhoods in (3.2a)–(3.2b), we look at the two vertices that are shared by the two tiles: the top one and the bottom one. For each of these 18 choices of a vertex z_0 we then find all possible ways to complete the picture to form a vertex neighborhood, keeping also in mind Remark 3.8.

E1, top vertex. Since along the edges with a single arrow we can only attach an L triangle, the tiling must continue as follows:

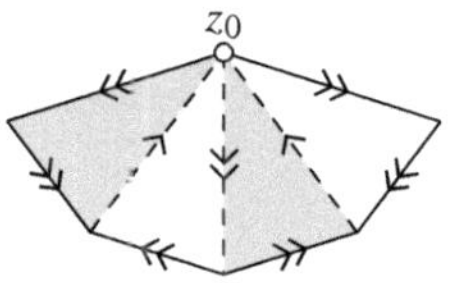

(3.8)

Next, we can attach to both the edges from z_0 an L triangle, which comes with another L triangle attached along the edge with a single arrow, and we get

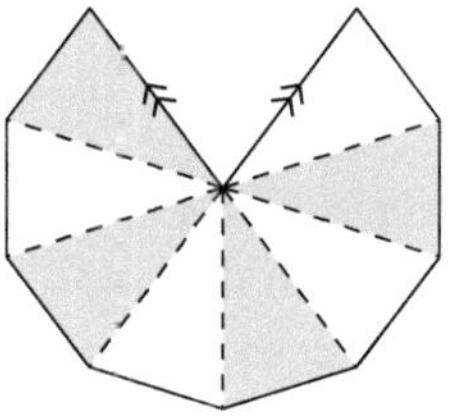

Now the only way to fill the remaining slot respecting the matching rule is with two L triangles, thus getting then the WHEEL.

Another way to continue (3.8) is by attaching to both edges starting from z_0 an S triangle. But each S triangle comes with an L triangle attached to the short edge with a double arrow (Remark 3.8), and we get the ICE CREAM (rotated by 180°).

Finally we can attach to an edge (for example the left one) an L triangle and to the other (the right one) an S triangle, and again because of Remark 3.8 we get

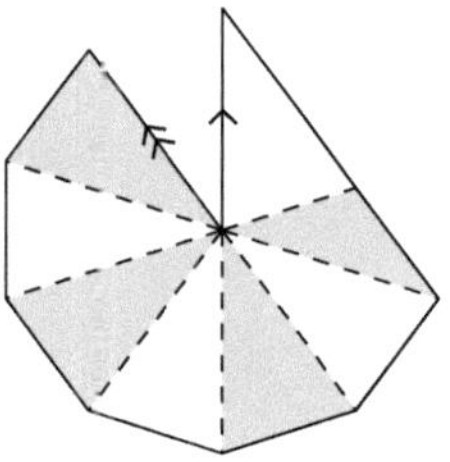

and there is no way to fill the empty space respecting the matching rule.

E1, bottom vertex. If we attach two S triangles along the short edges with a double arrow, we get the DIAMOND. If we attach two L triangles, we get

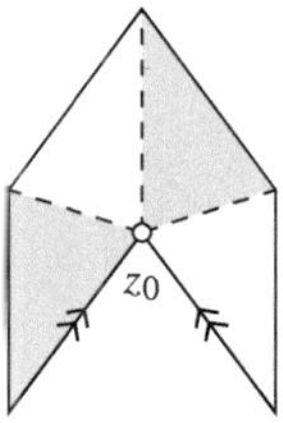

and the only way to finish the picture is with two S triangles, thus getting the SHIELD (rotated by 180°). Remark 3.8(2) tells us that these are the only two possibilities.

E2, top vertex. If we attach to the left or to the right an L triangle, this carries another L triangle attached along the edge with a single arrow, and we are back to the configuration (3.8). To get something new we must attach both sides an S triangle, and each carries an L triangle attached along the short edge with a double arrow. We get

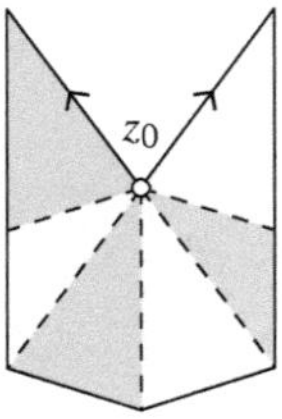

(3.9)

Now, each L triangle on top should have another L triangle attached along the edge with a single arrow, but the angles don't match.

E2, bottom vertex. If we attach to both short edges with a double arrow an L triangle, we get

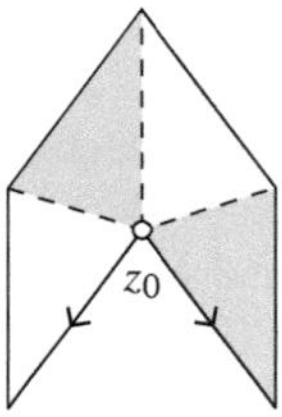

and we have the same problem that we had in (3.9).

If we attach to one edge (say the left one) an S triangle and to the other (the right one) an L triangle, the latter carries another L triangle attached along the edge with a single arrow, and we get:

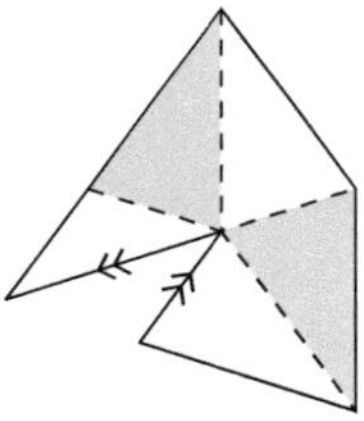

The only way to complete the picture is with an S triangle, getting the BIG KITE.

Finally if we attach to both edges an S triangle, we get:

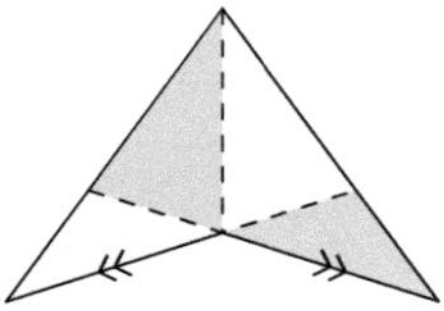

(3.10)

If we further attach an S triangle to one of the edges with a double arrow (say the one on the right), this carries an L triangle attached along the short edge with a double arrow, and we get:

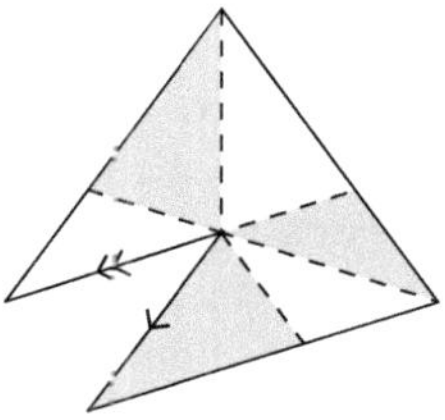

Now, the only thing we can attach along the edge with a single arrow is an L triangle, but the angle doesn't fit. The only possibility, then, is to attach to both edges in (3.10) an L triangle. Each will carry another L triangle attached along the edge with a single arrow, and we get the ice cream again.

E3, top vertex. If we attach to the edges starting from z_0 two L triangles, then there is no way to fill the remaining gap:

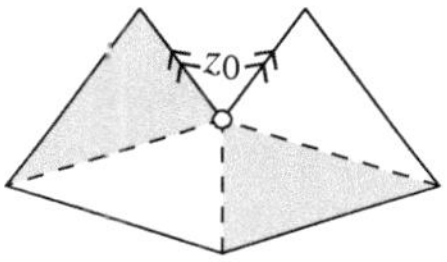

(3.11)

If we attach one L triangle and one S triangle, the latter carries another S triangle attached along the edge with a single arrow, and the only way to complete the picture is to get a shield. If we attach two S triangles, each will carry another S triangle attached along the edge with a single arrow, and we get:

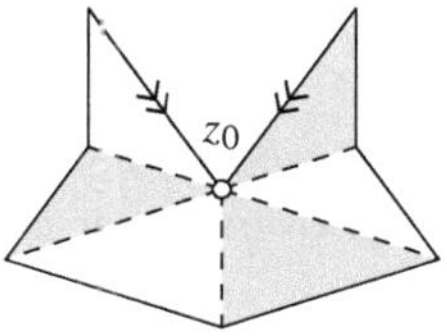

The only way to complete the picture is with two S triangles, and we get the CAT.

E3, bottom vertex. Along the edges with a single arrow we can only attach an L triangle, and we find a picture similar to (3.11) but with double arrows going in the opposite direction.

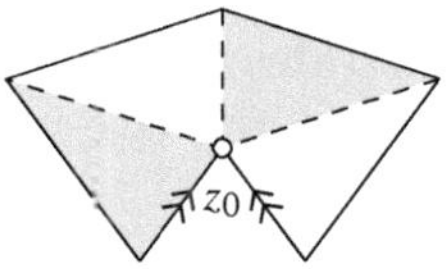

Now it is possible to complete the picture, and there is only one way to do it: with two S triangles. We get a big kite again.

E4, top vertex. We can only attach L triangles to the short edges with a double arrow. We get the diamond.

The strategy now should be clear. Let us only sketch the rest of the verification.

E4, bottom vertex. We get the cat, the shield or the STAR. This is the only case that requires a little bit of work. **E5, top vertex.** To the short edges with a double arrow we can only attach L triangles, that carry each another L triangle attached along the edge with a single arrow. We get a big kite. **E5, bottom vertex.** An S triangle has always another S triangle attached to the edge with a single arrow to form a kite. This is then a special case of E4, bottom vertex. We don't get any new vertex neighborhood (in fact, one can verify that in this case we only get the star and the cat). **E6, top vertex.** The L triangle always has another L triangle attached along the edge with a single arrow, and we get a special case of E2, bottom vertex. **E6, bottom vertex.** The S triangle has another S triangle attached along the edge with a single arrow, and the latter has an L triangle attached along the short edge with a double arrow. We get the diamond. **E7, top vertex.** The L triangle always has another L triangle attached along the edge with a single arrow, and we get a special case of E2, bottom vertex. **E7, bottom vertex.** The same as E6, bottom vertex. **E8, top vertex.** The S triangle always has another S triangle attached along the edge with a single arrow, and we get a special case of E4, bottom vertex. **E8, bottom vertex.** The L triangle always has another L triangle attached along the edge with a single arrow, and we get a special case of E2, top vertex. **E9, top vertex.** The same as E8, top vertex. **E9, bottom vertex.** The same as E8, bottom vertex. ■

The proof of Proposition 3.9 is a pedantic check of all possible ways (finitely many) of attaching tiles to get a vertex neighborhood. However, it has an interesting byproduct: along the way we meet a number of illegal partial tilings. We now derive some corollaries of the above classification that will be useful later on.

Corollary 3.10 *In a Robinson tiling:*

(i) every S triangle is contained in a diamond;
(ii) two distinct diamonds have non-overlapping interiors.

Proof To prove point (i), look at all possible vertex neighborhoods of the top vertex of an S triangle (the vertex of the isosceles triangle that is opposite to the base). The diamond is the only possibility. Point (ii) is obvious: if two diamonds share, say, and S triangle, they both must contain in their interior the top vertex of such a tile, hence they must coincide. The reasoning for L triangles is similar. ■

Proposition 3.11 *Every vertex neighborhood appears in the Cartwheel. More precisely, every vertex neighborhood appears in $\mathcal{C}_5$.*

Proof The patch $\mathcal{C}_5$ is in Fig. 3.6. One can spot all seven vertex neighborhoods in the picture. ■

Lemma 3.12 *In a Robinson tiling, every diamond is contained in an ace.*

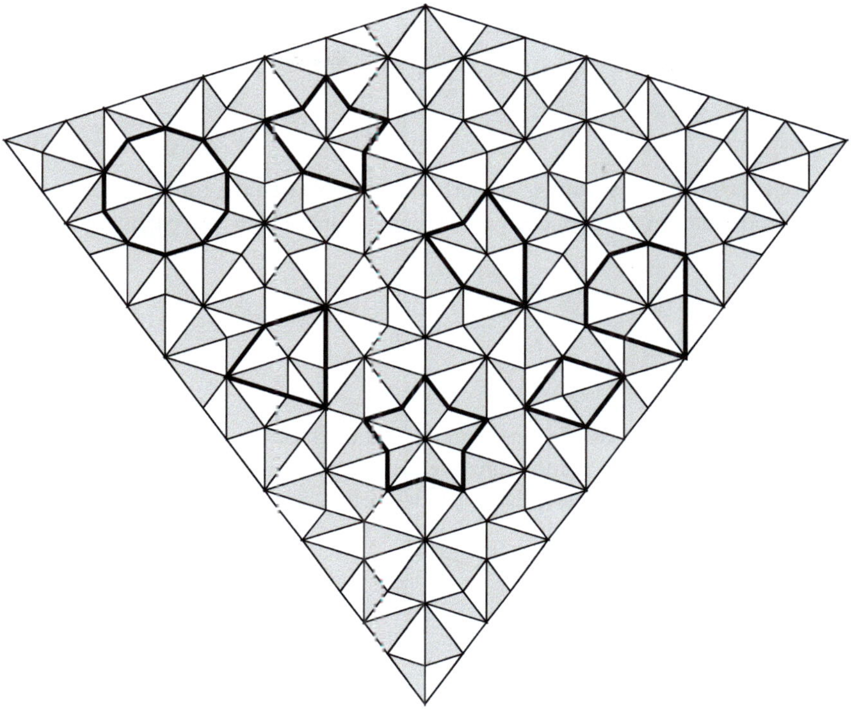

Fig. 3.6 The Cartwheel of order 5

Proof In a diamond,

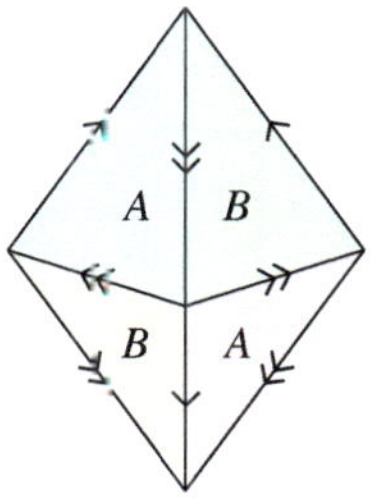

every L triangle has another L triangle attached along the edge with a single arrow, and we get an ace. ■

Proposition 3.13 *In a Robinson tiling, every tile is contained in an ace.*

Proof For S triangles this is trivial: every S triangle is contained in a diamond (Corollary 3.10), and every diamond is contained an ace (Lemma 3.12). We now pass to L triangles.

Every L triangle has another L triangle attached along the edge with a single arrow, i.e. is contained in a kite:

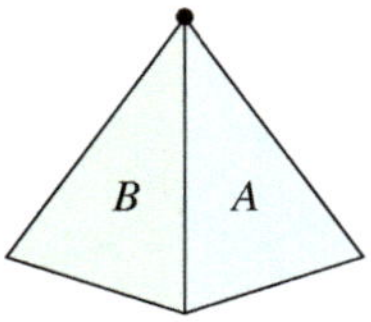

If we look at the list of vertex neighborhoods, we see that the top vertex of this kite is contained either in a wheel or in an ice cream. In both cases we get a patch:

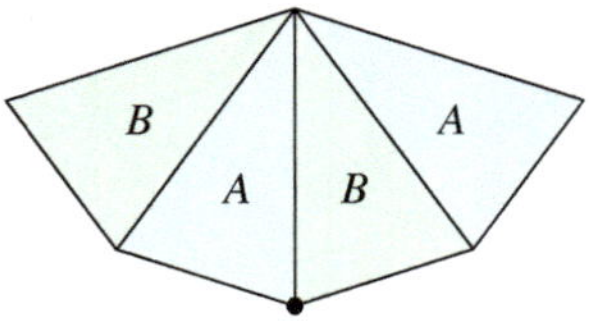

Next, we look at the possible vertex neighborhoods of the bottom vertex, marked with a bullet in the above picture, and we see that it is either a diamond or a shield. In the first case, if we complete the figure we get an ace. In the second case we get:

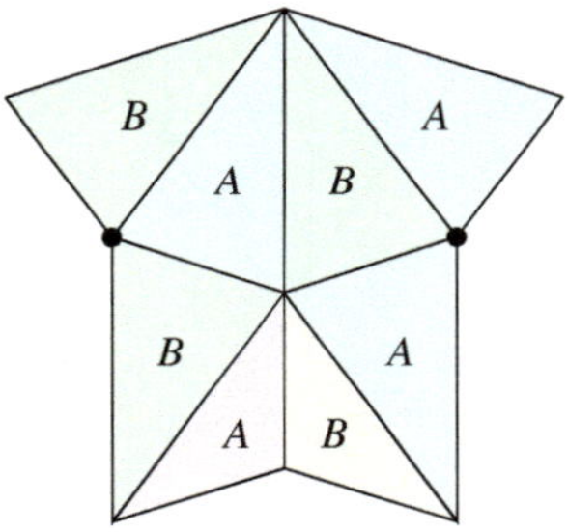

Finally, we look at the vertex neighborhoods of the two marked vertices, and we see that the only possibility is a big kite. Thus, we get:

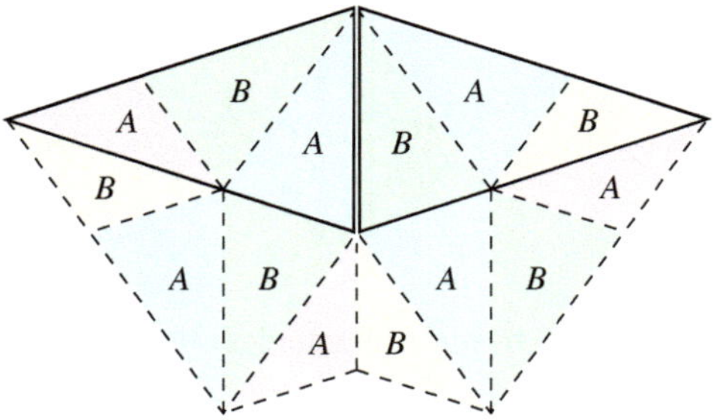

In the patch above, the framed triangles are each contained in an ace (since their S triangles are contained in an ace). This proves that the original L triangle, contained in one of the framed triangles in the picture, is in an ace as well. ■

It is interesting to see how vertex neighborhoods change under Decomposition. By a direct check one can verify that they change according to the following scheme:

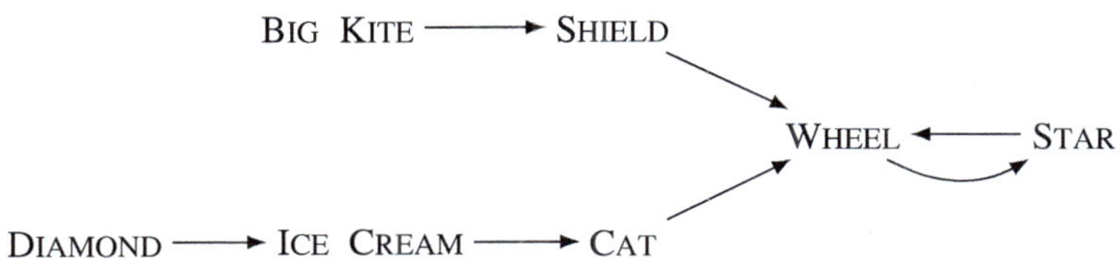

(3.12)

In particular, we see that after enough Decompositions every vertex neighborhood starts oscillating between a wheel and a star (but of course, new vertices are created at each Decomposition).

3.3 Composition and Aperiodicity

3.3.1 Composition Rules

Let $\mathfrak{P}$ be the protoset (3.1) and $\lambda > 0$. Decomposition is well-defined on any partial tiling with protoset $\lambda\mathfrak{P}$ covering a set $X \subseteq \mathbb{C}$ and produces a partial tiling of the same set X with protoset $\varphi^{-1}\lambda\mathfrak{P}$ (the prototiles get smaller, rescaled by a factor φ^{-1}). The opposite map (right to left in Fig. 3.1), enlarging the size of prototiles by removing arrows, is called

Composition

and is well-defined on tilings of the plane, as we shall prove in this section. Since Decomposition commutes with scalings, it is enough to prove it for $\lambda = 1$.

For starters we observe that, since in a tiling of the plane two distinct diamonds never overlap (Corollary 3.10), the following substitution rule is well-defined:

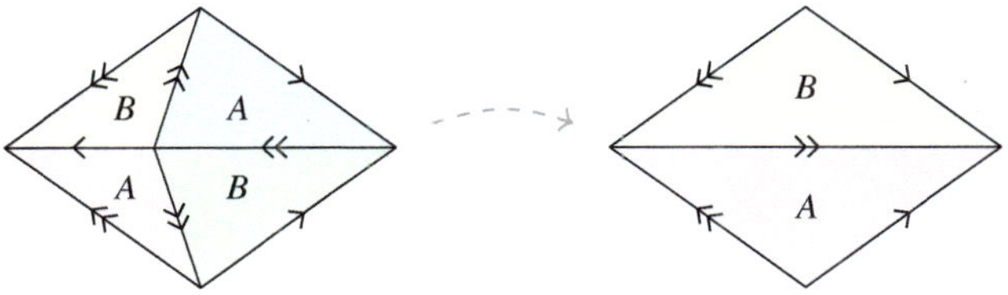

(3.13)

It takes every diamond and removes/replaces some of the internal edges. Since we are not changing the external edges, it transforms any tiling of the plane with protoset (3.1) into another tiling with protoset $\{L_A, L_B, S_A, S_B, \varphi S_A, \varphi S_B\}$ still respecting the matching rule. In fact, since in a Robinson tiling of the plane every S triangle is contained in a diamond, after the substitution (3.13) we get rid of all S triangles and obtain a tiling of the plane with protoset:

$$\{L_A, L_B, \varphi S_A, \varphi S_B\}. \tag{3.14}$$

The substitution rule (3.16) will be called *Composition 1*.

Lemma 3.14 *In a tiling of the plane with protoset* (3.14)*, every L_A tile (resp. L_B) has always a φS_B tile (resp. φS_A) attached along the edge with a single arrow.*

Proof This is similar to the proof of Lemma 3.7. Suppose a tile L_A has a tile L_B attached along the edge with a single arrow. Both tiles must have an L triangle attached along the short edge with a double arrow (we cannot attach a φS tile because the length of the edges do not match). Hence, we have a configuration like this:

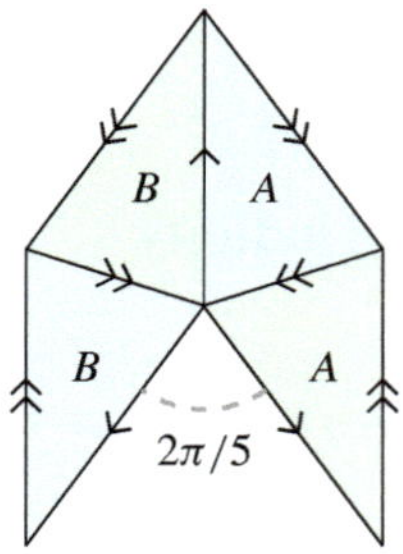

But there are no prototiles in (3.14) that can fill the empty white space respecting the matching rules. ■

Lemma 3.15 *In a tiling of the plane with protoset* (3.14)*: (i) every L triangle is contained in a big kite:*

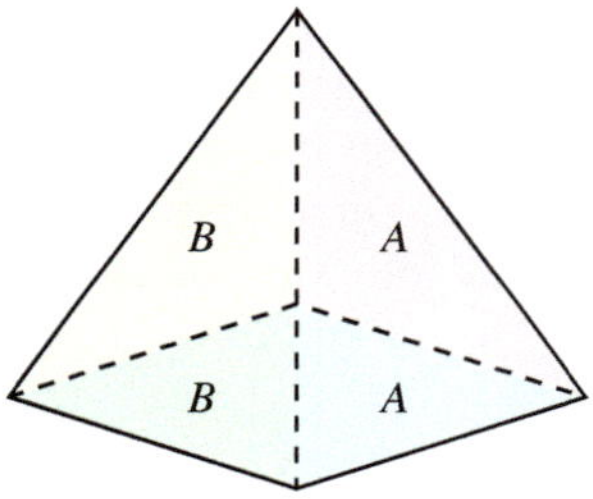

(ii) two distinct big kites have non-overlapping interiors.

Proof This is similar to Corollary 3.10. From Lemma 3.14 it follows that, in a tiling of the plane with protoset (3.14), L triangles always come in combinations:

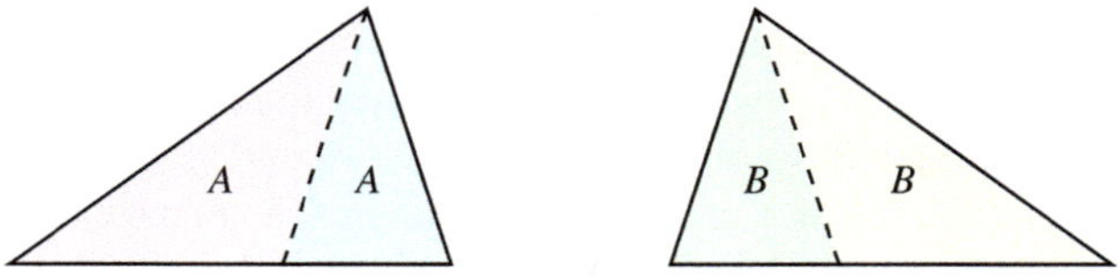

Since in the protoset (3.14) the only edges of length 1 are the bases of the L triangles (the other edges being of length φ or $\varphi + 1$), the two patches above always come in pairs, attached to each other to form a big kite. Two big kites cannot overlap. ■

Therefore, there is a well-defined map transforming tilings of the plane with protoset (3.14) into a tiling with protoset

$$\{\varphi L_A, \varphi L_B, \varphi S_A, \varphi S_B\}, \tag{3.15}$$

given by the substitution

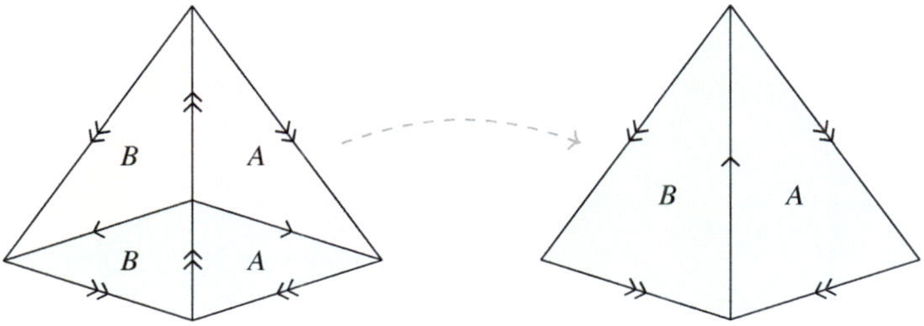

(3.16)

This map takes every big kite and removes/replaces some of the internal edges. Again, since we are not changing the external edges, we get a tiling still respecting the matching rule. Since every L triangle is contained in a big kite, after this substitution all L triangles are gone, and we remain with a tiling with protoset (3.15).

The substitution in (3.16) will be called *Composition 2*. Applying Composition 1 followed by Composition 2, we can replace any tiling of the plane with protoset (3.1) by one with protoset (3.15). This is what we call *Composition*.

For tilings of the plane, we have then the following commutative diagram:

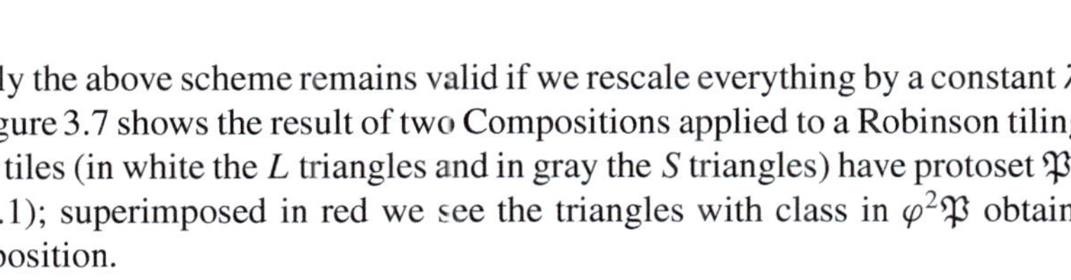

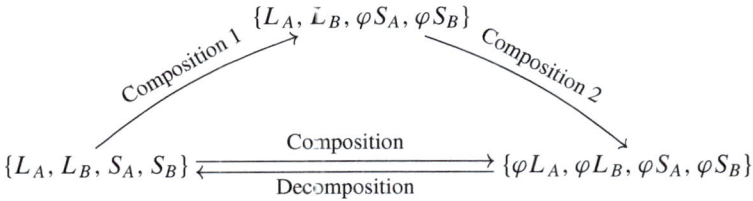

Clearly the above scheme remains valid if we rescale everything by a constant $\lambda > 0$.

Figure 3.7 shows the result of two Compositions applied to a Robinson tiling. The small tiles (in white the L triangles and in gray the S triangles) have protoset $\mathfrak{P}$ given by (3.1); superimposed in red we see the triangles with class in $\varphi^2\mathfrak{P}$ obtained by Composition.

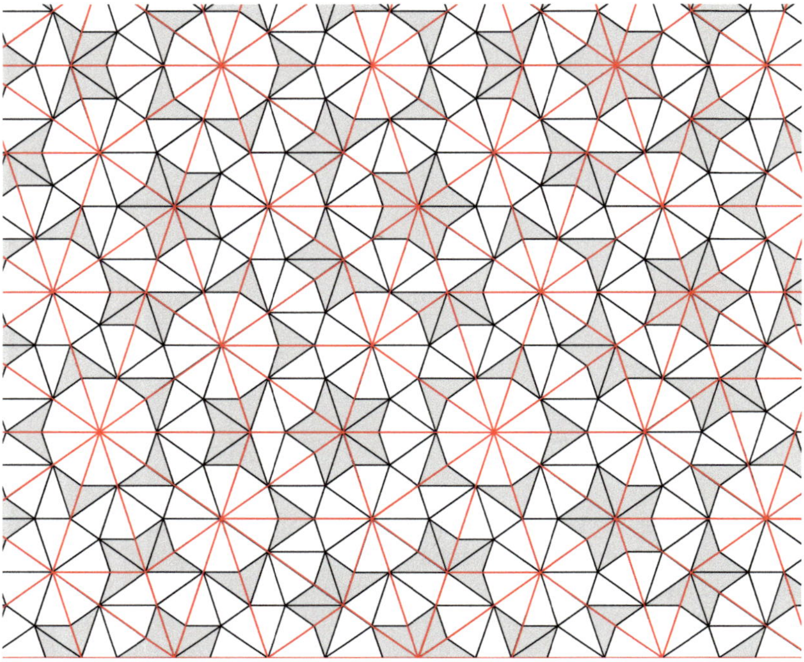

Fig. 3.7 Hierarchical structure of a Robinson tiling

3.3.2 *Aperiodicity of Robinson Triangles*

Using Composition we now show that Robinson triangles can only tile the plane non-periodically.

Proposition 3.16 *The protoset (3.1) is aperiodic.*

Proof Let $\mathfrak{P}$ be a protoset with at least one bounded prototile. Recall that there exists $\delta > 0$ such that any translation of a non-zero vector with norm $< \delta$ is not a symmetry of any tiling with prototile $\mathfrak{P}$ (Lemma 2.11). We can choose any δ smaller than the inradius of all the prototiles in $\mathfrak{P}$. When $\mathfrak{P}$ is the protoset (3.1), we can choose for example $\delta = 1/4$ (cf. Appendix A.1).

Now, let $\mathcal{T}_0$ be a tiling of the plane with protoset (3.1) and assume, by contradiction, that $\mathcal{T}_0 = \mathcal{T}_0 + c$ for some $c \in \mathbb{C} \setminus \{0\}$.

If we apply Composition n times we get a tiling $\mathcal{T}_n$ with protoset $\varphi^n \mathfrak{P}$. Since Composition commutes with translations, $\mathcal{T}_n = \mathcal{T}_n + c$ (observe that we are not rescaling the tilings, so the translation parameter c is unchanged). From the discussion above it follows that the translation $z \mapsto z + c$ cannot be a symmetry of $\mathcal{T}_n$ if $|c| < \varphi^n \delta$. Since $\varphi^n \to \infty$ for $n \to \infty$, for n big enough we get a contradiction. ■

One may wonder why the proof of the above proposition does not work for edge-to-edge tilings by unit squares (that are bi-periodic). There is an obvious subdivision

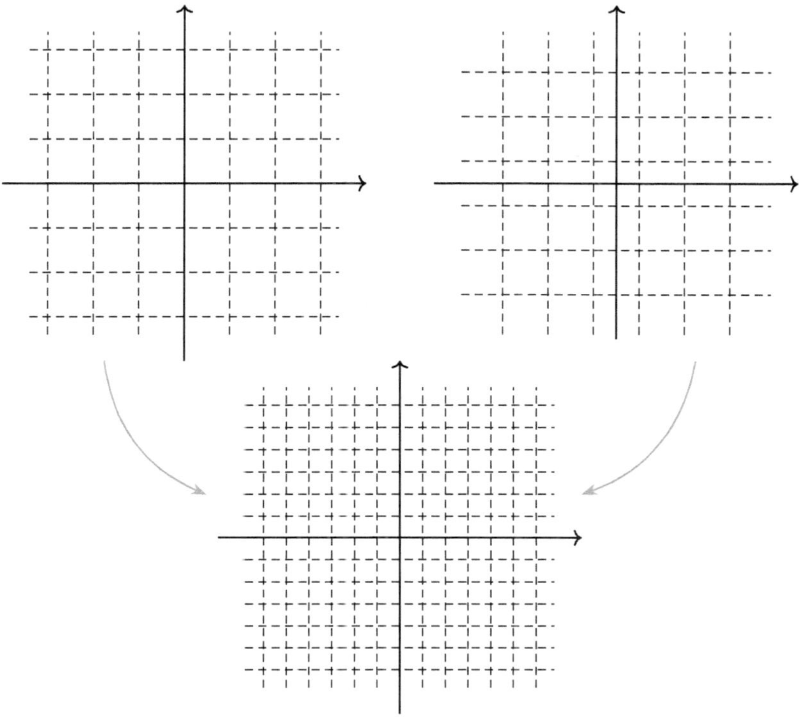

Fig. 3.8 Subdivision of square tilings is not injective

rule for square tilings, which allows to pass from a tiling with squares of size $\lambda > 0$ to a tiling with squares of size $\lambda/2$:

However, this map is not injective, so an inverse map does not exist. In Fig. 3.8, we see an example of two square tilings that become the same after subdivision.

3.3.3 Local Properties

In this section we shall prove that Robinson tilings are locally indistinguishable. Here $\mathcal{C}_n$ is the Cartwheel of order n described in Sect. 3.1.1. Recall that, if $\mathcal{T}$ is a tiling and $\mathcal{P} \subset \mathcal{T}$ a patch, by a *copy* of $\mathcal{P}$ we mean another patch $\mathcal{P}'$ (in the same or in a different tiling) that can be transformed into $\mathcal{P}$ with a direct isometry.

Proposition 3.17 *In a Robinson tiling, and for all $n \geq 1$, every tile is contained in a copy of $\mathcal{C}_n$.*

Proof Start with n odd. Let $\mathcal{T}$ be the tiling and $T_0 \in \mathcal{T}$ any tile. Apply Composition $n-1$ times and get a tiling $\mathcal{T}'$ with tiles $\varphi^{n-1}L$ and $\varphi^{n-1}S$. Now, T_0 is contained in a tile of $\mathcal{T}'$ and the latter is contained in an ace made with these huge triangles (cf. Proposition 3.13). If we apply Decomposition $n-1$ times to this big ace we get a Cartwheel $\mathcal{C}_n$ contained in $\mathcal{T}$ and containing T_0. The proof for n even is similar and we omit it. ■

Corollary 3.18 *Every Robinson tiling contains a copy of $\mathcal{C}_n$, and in fact infinitely many distinct copies, for every $n \geq 1$.*

Proof It follows from Proposition 3.17 that any Robinson tiling $\mathcal{T}$ contains a copy of $\mathcal{C}_n$. By contradiction, suppose that there are only finitely many distinct copies of $\mathcal{C}_n$ in $\mathcal{T}$. Since $\mathcal{C}_n$ is bounded, it must exist a tile in $\mathcal{T}$ that is not contained in any of these copies, thus contradicting Proposition 3.17. ■

Corollary 3.19 *In a Robinson tiling, every vertex neighborhood appears infinitely many times.*

Proof It follows from Corollary 3.18 and Proposition 3.11. ■

Lemma 3.20 *In a Robinson tiling, for all $z_0 \in \mathbb{C}$ and all $0 < r < 1/4$, the disk $D_r(z_0)$ is contained in a vertex neighborhood $\mathcal{N}$ (by this we mean that $D_r(z_0) \subset \bigcup_{T \in \mathcal{N}} T$).*

Proof Since $r < 1/4$ is less than the radius of the inscribed circle of both the L and S prototiles (cf. Appendix A.1), the disk $D_r(z_0)$ can intersect at most two edges of a tile. If $D_r(z_0)$ is in the interior of a tile, then it is contained in a vertex neighborhood. If $D_r(z_0)$ contains a vertex v then, called $\mathcal{N}$ the vertex neighborhood of v, for every $T \in \mathcal{T}$ the disk intersects the two edges incident to v, which means that it cannot intersect the third edge: the disk is then contained in $\mathcal{N}$. These two cases are illustrated in the next pictures:

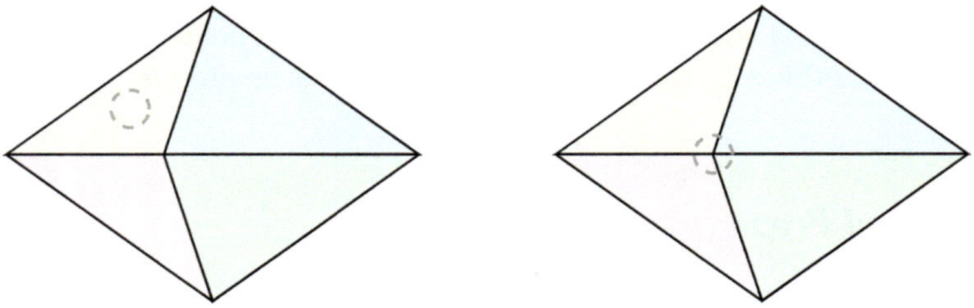

If $D_r(z_0)$ intersects at most one edge of a tile, then it is contained in an edge neighborhood, like for example in the next picture:

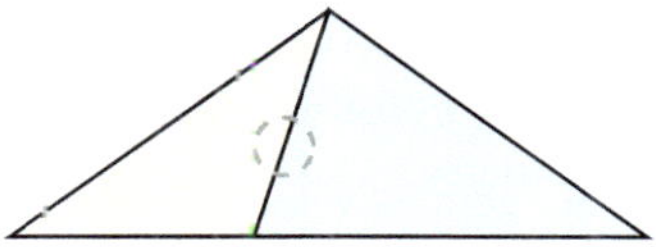

Finally, suppose $D_r(z_0)$ intersects two edges of some tile T but does not contain any vertex, like for example in the next picture where T is a golden gnomon:

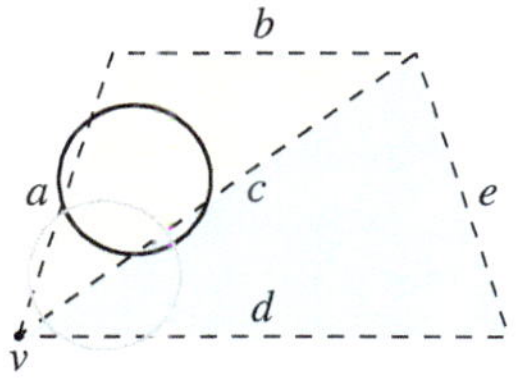

Here both the black and the gray disks intersect the edges a and c of the S tile. Since r is smaller than the inner radius of the tile, any disk $D_r(z_0)$ intersecting the edges a and c, sharing the vertex v, cannot intersect the edge b, opposite to v. Now we look at the tile attached to c, an L triangle in the picture. A disk of radius $r < 1/4$ crossing the edges a and c could cross the edge d, the one incident with v, but not the opposite one. Indeed, the distance between any two points in the edges a and e is greater than 1 (the length of b), and then greater than the diameter $2r$ of the disk. Any disk of radius $r < 1/4$ intersecting a cannot intersect e.

We repeat the reasoning for all possible edge neighborhoods in (3.2a)–(3.2b), say formed by two adjacent tiles T_1 and T_2, and show that if a disk of radius $r < 1/4$ crosses two edges of T_1, called v the common vertex to these two edges, then the disk cannot intersect the edge of T_2 opposite of v. Finally, we prove by induction that if $D_r(z_0)$ intersects two edges of a tile, sharing a vertex v, then any other edge intersecting the disk must be incident with (and never opposite to) the vertex v. This proves that $D_r(z_0)$ is contained in the neighborhood of v. ■

Proposition 3.21 *Let $\mathcal{T}$ and $\mathcal{T}'$ be two Robinson tilings and $\mathcal{P} \subset \mathcal{T}$ a patch. Then, $\mathcal{T}'$ contains infinitely many copies of $\mathcal{P}$.*

Proof Call $\mathcal{T}_n$ (resp. $\mathcal{T}_n'$) the tiling obtained by applying Composition n times to $\mathcal{T}$ (resp. $\mathcal{T}'$). The patch $\mathcal{P}$ is contained in a disk $D_r(z_0)$ for some z_0 and r. It follows from Lemma 3.20 that, for n large enough, the disk $D_r(z_0)$ (and then the patch) is contained in a vertex neighborhood $\mathcal{N}$ of $\mathcal{T}_n$ (more precisely, this is true for every n such that $r < \varphi^n/4$).

There is a copy $\mathcal{N}'$ of the same vertex neighborhood in $\mathcal{T}_n'$ (Corollary 3.19). If we apply Decomposition n times to the two vertex neighborhoods we obtain congruent patches $\mathcal{P}_0$ and $\mathcal{P}_0'$ in $\mathcal{T}$ and $\mathcal{T}'$, respectively. Since $\mathcal{P}_0$ contains $\mathcal{P}$, it follows that $\mathcal{P}_0'$ contains a copy of $\mathcal{P}$.

In fact, since there are infinitely many copies of $\mathcal{N}'$ in $\mathcal{T}_n'$ (Corollary 3.19), there are infinitely many copies of $\mathcal{P}$ in $\mathcal{T}'$. ■

A consequence of the previous proposition is that in any Robinson tiling one can find infinitely many and arbitrarily large patches with 5-fold symmetry, since the tiling contains infinitely many arbitrarily large patches of both the Sun and the Star tiling. On the other hand, the next proposition shows that almost all Robinson tilings do not have a global 5-fold symmetry.

Proposition 3.22 *The Sun and the Star tilings of Sect. 3.1.2 are (up to a congruence) the only Robinson tilings with 5-fold symmetry.*

Proof Let $\mathcal{T}$ be a Robinson tiling with 5-fold symmetry. The center of symmetry in a non-periodic tiling is unique (Lemma 2.10(i)), and up to a translation we can assume that it is the origin of the plane. The vertex neighborhood of 0 must be a wheel or a star, since they are the only ones with 5-fold symmetry. Let us assume that it is a wheel, the other case being similar.

Apply Composition n times and call $\mathcal{T}^{(n)}$ the corresponding tiling of the plane. Since Composition and Decomposition commute with isometries, $\mathcal{T}^{(n)}$ has also 5-fold symmetry with center of symmetry 0. The vertex neighborhood of 0 in $\mathcal{T}^{(n)}$ must be a huge wheel or a huge star, for the same reason as above. Call $\mathcal{P}_n \subset \mathcal{T}$ the patch obtained from $\mathcal{T}^{(n)}$ by applying Decomposition n times to the above-mentioned vertex neighborhood. This is a patch in the Sun tiling by the above assumption (the neighborhood of 0 in $\mathcal{T}$ is a wheel; if the neighborhood in $\mathcal{T}^{(n)}$ is also a wheel, then n must be even; if it is a star, then n must be odd).

We have $\mathcal{P}_n \subset \mathcal{P}_{n+1}$, since they are patches in the same tiling $\mathcal{T}$. Since

$$\mathcal{T}' := \bigcup_{n \geq 0} \mathcal{P}_n$$

covers the plane, it must be $\mathcal{T}' = \mathcal{T}$. But $\mathcal{T}'$ is the Sun tiling. ■

It follows from Lemma 3.20 that, if x_0, y_0 are any two points of a Robinson tiling, after applying Composition enough times the two points will belong to the same vertex neighborhood. Not necessarily to the same tile, though. For example, in a Sun or Star tiling, if x_0 and y_0 are antipodal points in the vertex neighborhood of the center of symmetry, after an arbitrary number of Compositions they will still belong to different tiles of a wheel or star neighborhood. The above claim, nevertheless, can be improved a little bit. Firstly, in (3.12) we saw how vertex neighborhoods change under Decomposition. Similarly, we can also study how vertex neighborhoods change under Composition, with the difference that the result of Composition applied to a patch may depend on the tiles surrounding that patch.

Proposition 3.23 *Let z_0 be a vertex in a Robinson tiling $\mathcal{T}$ and $\mathcal{N}$ its vertex neighborhood. Let $\mathcal{T}'$ be the tiling obtained by applying Composition to $\mathcal{T}$. If $\mathcal{N}$ is a diamond or a big kite, then z_0 is not a vertex in $\mathcal{T}'$. In every other case, the vertex neighborhood of z_0 changes according to the following diagram:*

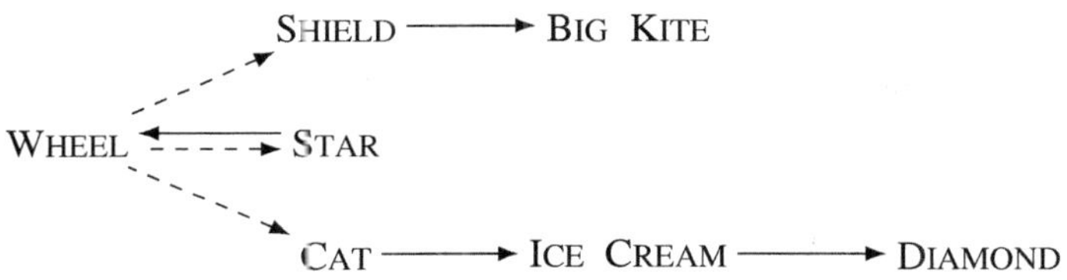

(Observe that $\mathcal{N}'$ is uniquely determined by $\mathcal{N}$ except when $\mathcal{N}$ is a wheel.)

Proof Let $\mathcal{N}'$ be the neighborhood of z_0 in $\mathcal{T}'$. In general, applying Composition to $\mathcal{T}$, some of the edges incident with z_0 might disappear. If they all disappear, then $\mathcal{N}'$ is a single tile (and z_0 a point in its interior). If only one edge remains (in fact, two edges incident with z_0 and with the same direction), then $\mathcal{N}'$ is a patch of two tiles, i.e. one of the nine edge neighborhoods in (3.2a)–(3.2b), and z_0 is a point on the common edge between the two tiles. In all other cases, z_0 is a vertex of $\mathcal{T}'$ and $\mathcal{N}'$ is a vertex neighborhood.

Observe that if we apply Decomposition to a single tile we do not get any vertex in the interior of the tile, and if we apply Decomposition to one of the nine edge neighborhoods in (3.2a)–(3.2b), one checks case by case that either we do not get any new vertex, or we get a new vertex whose neighborhood is a big kite or a diamond. See, for example, (3.4).

Assume that $\mathcal{N}$ is not a big kite or a diamond. By the argument above $\mathcal{N}'$ cannot be a single tile or an edge neighborhood. The only remaining possibility is that z_0 is a vertex in $\mathcal{T}'$ and $\mathcal{N}'$ is a vertex neighborhood. Since (3.12) tells us how vertex neighborhoods change under Decomposition, reversing the arrows we see how they change under Composition.

Finally, let $\mathcal{N}$ be a diamond or a big kite and assume that z_0 is a vertex in $\mathcal{T}'$. Then, $\mathcal{N}'$ is a vertex neighborhood which should become $\mathcal{N}$ after Decomposition. But there is no vertex neighborhood that becomes a big kite or a diamond after Decomposition, hence we get a contradiction: z_0 cannot be a vertex in $\mathcal{T}'$. ■

3.4 Index Sequences

Let $\mathcal{T}$ be a tiling of the plane with protoset (3.1) and $z_0 \in \mathbb{C}$ any point in the interior of a tile $T_0 \in \mathcal{T}$. Observe that T_0 is uniquely determined by z_0.

Set $\mathcal{T}_0 := \mathcal{T}$. By induction, if $n \geq 1$ is odd, let $\mathcal{T}_n$ be the tiling obtained from $\mathcal{T}_{n-1}$ applying Composition 1. If $n \geq 1$ is even, let $\mathcal{T}_n$ be the tiling obtained from $\mathcal{T}_{n-1}$ applying Composition 2. Thus, for every n, the tiling $\mathcal{T}_n$ is similar to one with protoset (3.1) if n is even, and is similar to one with protoset (3.14) if n is odd.

Clearly, every point that is in the interior of a tile, is still in the interior of a tile after Composition 1 or 2. Denote by T_n the (unique) tile of $\mathcal{T}_n$ containing z_0 in its interior. We now transform the sequence $(T_0, T_1, T_2, \dots)$ of tiles in a binary sequence

$$\iota(\mathcal{T}, z_0) = (a_n)_{n\in\mathbb{N}} \in \{0, 1\}^{\mathbb{N}}$$

as follows. For all even $n \geq 0$,

$$\text{if } T_n \text{ is a } \begin{matrix}\text{golden triangle}\\ \text{golden gnomon}\end{matrix} \text{ then } a_n = \begin{matrix}0\\1\end{matrix}.$$

For all odd $n \geq 0$,

$$\text{if } T_n \text{ is a } \begin{matrix}\text{golden triangle}\\ \text{golden gnomon}\end{matrix} \text{ then } a_n = \begin{matrix}1\\0\end{matrix}.$$

Note that the sequence is constructed using only the shape of the tiles, and not the type. We will see that one can reconstruct a tiling $\mathcal{T}$ from its index sequence up to a rotation, a translation and, this time, also a reflection.

Observe also that, in a tiling $\mathcal{T}_n$ with n even the golden triangle is the one with the bigger area, while in a tiling $\mathcal{T}_n$ with n odd the golden gnomon is the one with bigger area (cf. Appendix A.1). Thus, the rule above associates the digit 1 to the smaller triangle and the digit 0 to the larger triangle, regardless of the parity of n.

Definition 3.24 $\iota(\mathcal{T}, z_0)$ is called the *index sequence* of the pair $(\mathcal{T}, z_0)$.

As an exercise, we now compute the beginning of the index sequence of the tiling in Fig. 3.9 (clearly, from a patch one can compute only finitely many terms in the index sequence). The point z_0 is marked with a red cross in the picture.

The first tiling is $\mathcal{T}_0$; the next $\mathcal{T}_1$, which is obtained with Composition 1 removing some internal edges from every diamond; then $\mathcal{T}_2$, which is obtained with Composition 2 removing some internal edges from every big kite; etc. In each tiling: the thin gray lines are those that will disappear after the next Composition, while the thick edges remain; the color of each tile is white if the tile gives a 0 in the index sequence (i.e. the tile is large), gray if it gives a 1 (i.e. the tile is small). The index sequence is then $(0, 1, 0, 1, 0, 0, 1, 0, \ldots)$.

3.4.1 Some Periodic Index Sequences

The purpose of this section is to compute the index sequence of the tilings in Sect. 3.1 for a suitable choice of the base point z_0 (we will see later on that the tail of the index sequence is independent of z_0). We need the following lemma.

Lemma 3.25 *Let $\mathcal{T}$ be a Robinson tiling such that $\mathcal{F}^{n_0}(\mathcal{T}) = \mathcal{T}$ for some $n_0 \geq 1$. Let $T_0 \in \mathcal{T}$ be a tile containing* 0 *(not necessarily in its interior) and $z_0 \in \mathring{T}_0$. Then, $\iota(\mathcal{T}, z_0)$ is periodic with period $2n_0$.*

Proof Let $(a_k)_{k\in\mathbb{N}} := \iota(\mathcal{T}, z_0)$ be the index sequence. Set $\mathcal{T}_0 := \mathcal{T}$. Call $\mathcal{T}_k$ the tiling obtained from $\mathcal{T}_{k-1}$ by applying Composition 1 if $k \geq 1$ is odd, or Composition 2 if $k \geq 2$ is even. Let $T_k \in \mathcal{T}_k$ be the tile containing z_0 in its interior. Since our prototiles

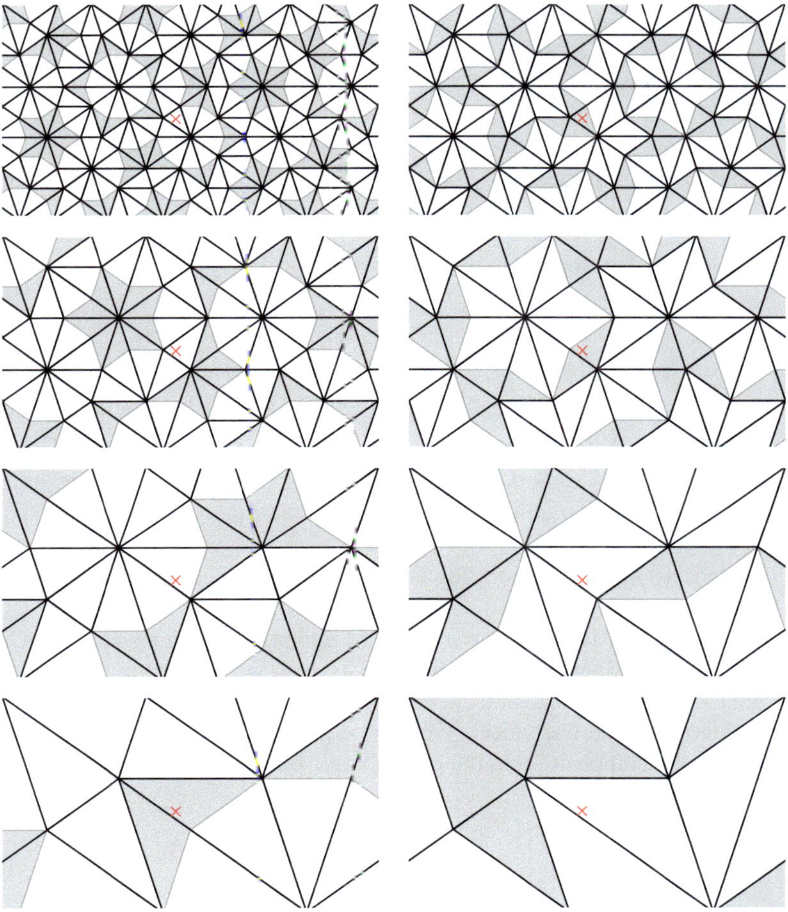

Fig. 3.9 Computing an index sequence

are convex, and $T_k \supset T_0$ contains 0, it contains tz_0 in its interior for all $0 < t \leq 1$. Observe that

$$\mathcal{F}(\mathcal{T}_k) = \varphi \mathcal{T}_{k-2} \qquad \forall\, k \geq 2 \tag{3.17}$$

(Decomposition is the inverse map of Composition 1 plus Composition 2, and $\mathcal{F}$ is Decomposition composed with a scaling of a factor φ.)

By assumption $\mathcal{F}^{n_0}(\mathcal{T}) = \mathcal{T}$, but $\mathcal{F}$ and Composition commute, hence $\mathcal{T}_k = \mathcal{F}^{n_0}(\mathcal{T}_k)$ for all even k. Combined with (3.17), this gives

$$\mathcal{T}_k = \varphi^{n_0} \mathcal{T}_{k-2n_0} \tag{3.18}$$

for all even $k \geq 2n_0$. Now, $T_{k-2n_0} \in \mathcal{T}_{k-2n_0}$ contains $\varphi^{-n_0} z_0$ in its interior, thus $\varphi^{n_0} T_{k-2n_0}$ contains z_0, which by (3.18) means that

$$\varphi^{n_0} T_{k-2n_0} = T_k.$$

We see that T_k is similar to T_{k-2n_0}, and since the shift in the sequence is even, this implies $a_k = a_{k-2n_0}$. A similar reasoning applies to odd k. ■

We now check that, for a suitable choice of z_0:

$\iota(\text{Cartwheel}, z_0) = (0, 0, 0, 0, 0, \ldots)$	(null sequence)
$\iota(\text{Star}, z_0) = (1, 0, 0, 0, 1, 0, 0, 0, 1, 0, 0, 0, \ldots)$	(1000 repeated indefinitely)
$\iota(\text{Golden Triangle}, z_0) = (0, 1, 0, 0, 0, 1, 0, 0, 0, \ldots)$	(like the Star, but shifted by 1 slot)
$\iota(\text{Sun}, z_0) = (0, 0, 1, 0, 0, 0, 1, 0, 0, 0, \ldots)$	(like the Star, but shifted by 2 slots)
$\iota(\text{Golden Gnomon}, z_0) = (0, 0, 0, 1, 0, 0, 0, 1, 0, 0, \ldots)$	(like the Star, but shifted by 3 slots)
$\iota(\text{Golden Gnomon}', z_0) = (1, 0, 1, 0, 1, 0, 1, 0, \ldots)$	(10 repeated indefinitely)

Here the *Golden Gnomon* is the tiling $\mathcal{G}$ and the *Golden Gnomon*$'$ is the tiling $\mathcal{G}'$ of Sect. 3.1.3.

The Cartwheel satisfies the hypothesis of Lemma 3.25 with $n_0 = 2$; the Golden Gnomon$'$ with $n_0 = 5$; every other tiling with $n_0 = 4$. We must compute the first $2n_0$ terms in each index sequence and check that we get what expected.

Let us start with the Cartwheel. We choose any z_0 in the edge neighborhood of the origin, apply Composition to the patch $\mathcal{C}_3$ and obtain:

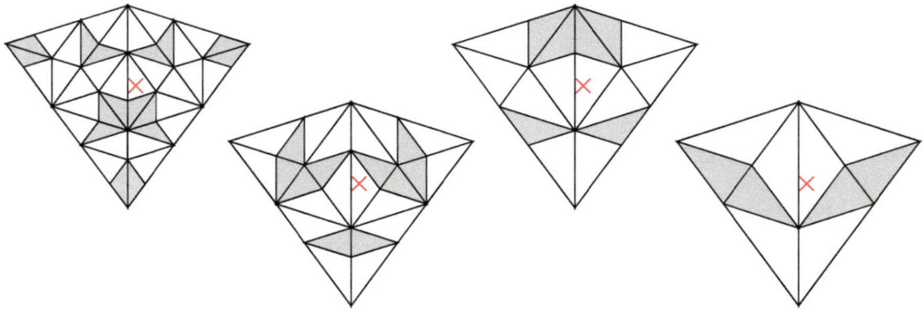

We see that the index sequence starts with four zeros.

Next, the Sun. We must start with the patch $\mathcal{T}_4$. We choose z_0 like in the next picture and see that the first eight terms in the index sequence are the expected ones.

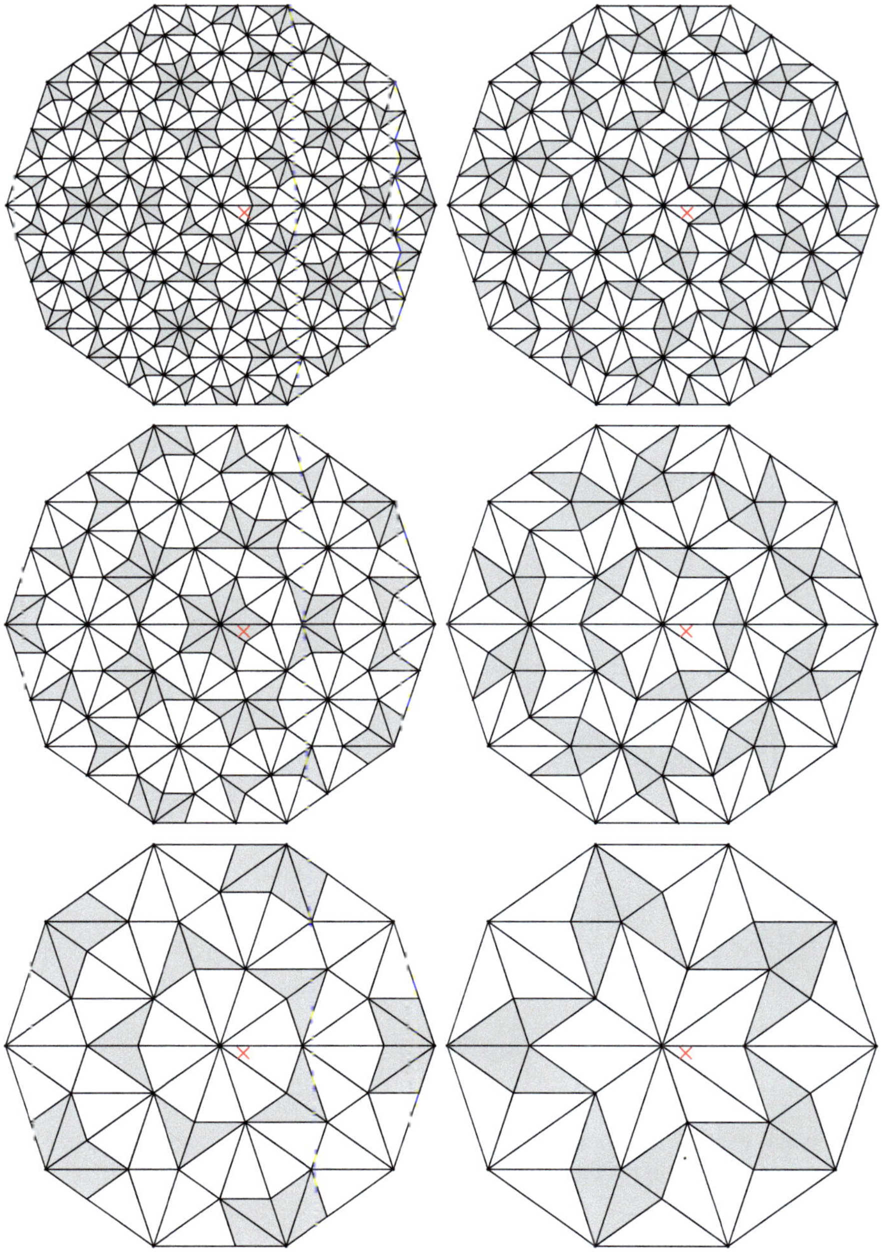

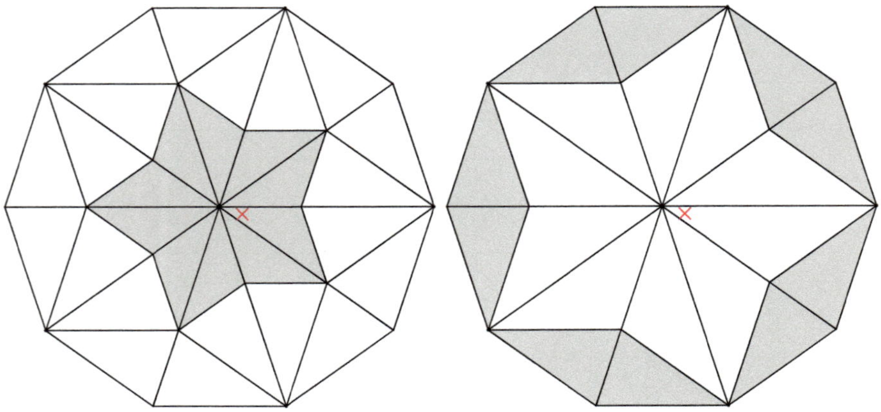

For the next three tilings, we choose $z_0 = 0$. We first compute the first eight terms of the index sequence of the Golden Triangle. We start with the patch $\mathcal{T}_1$.

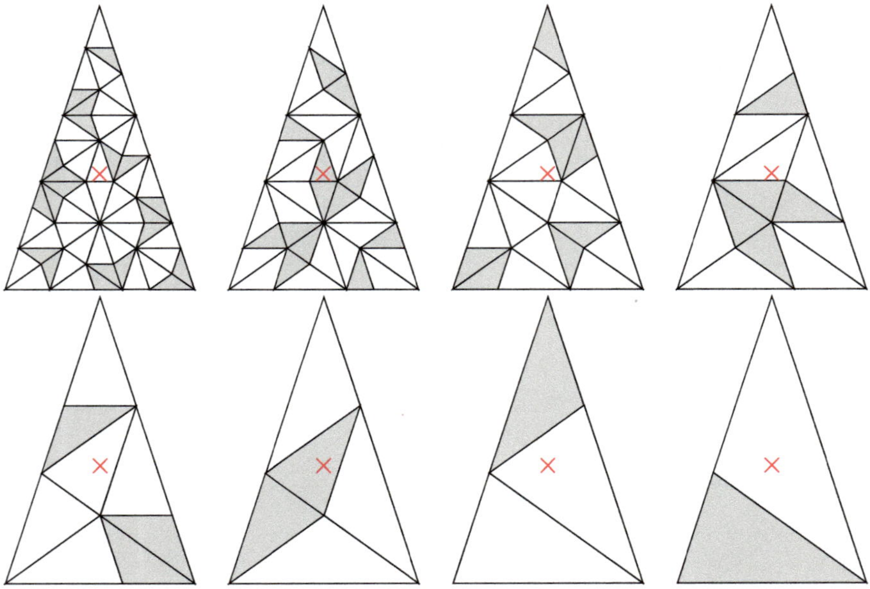

We see that the first eight terms in the index sequence are the expected ones. We do the same computation for the Golden Gnomon, starting with the patch $\mathcal{G}_1$.

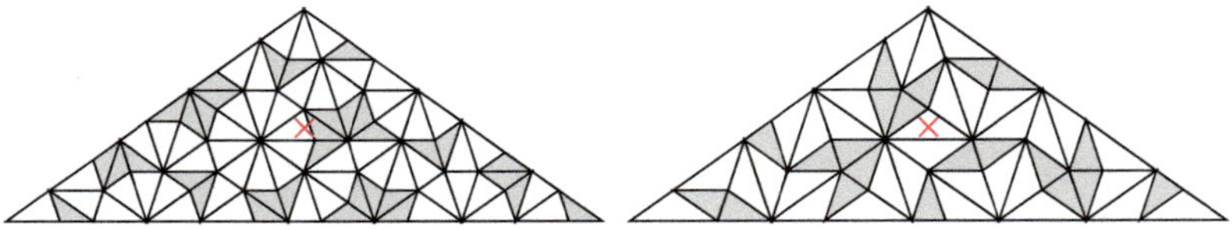

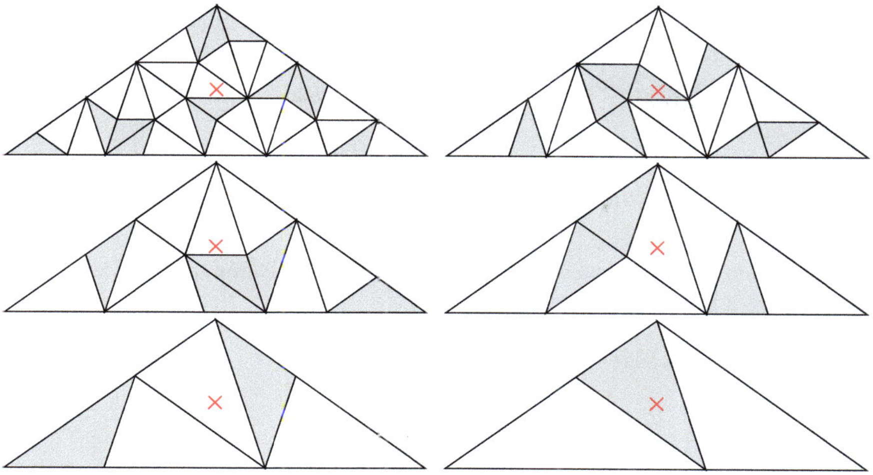

Next, we compute the first eight terms of the index sequence of the Golden Gnomon′. We start with the patch $\mathcal{G}_1'$.

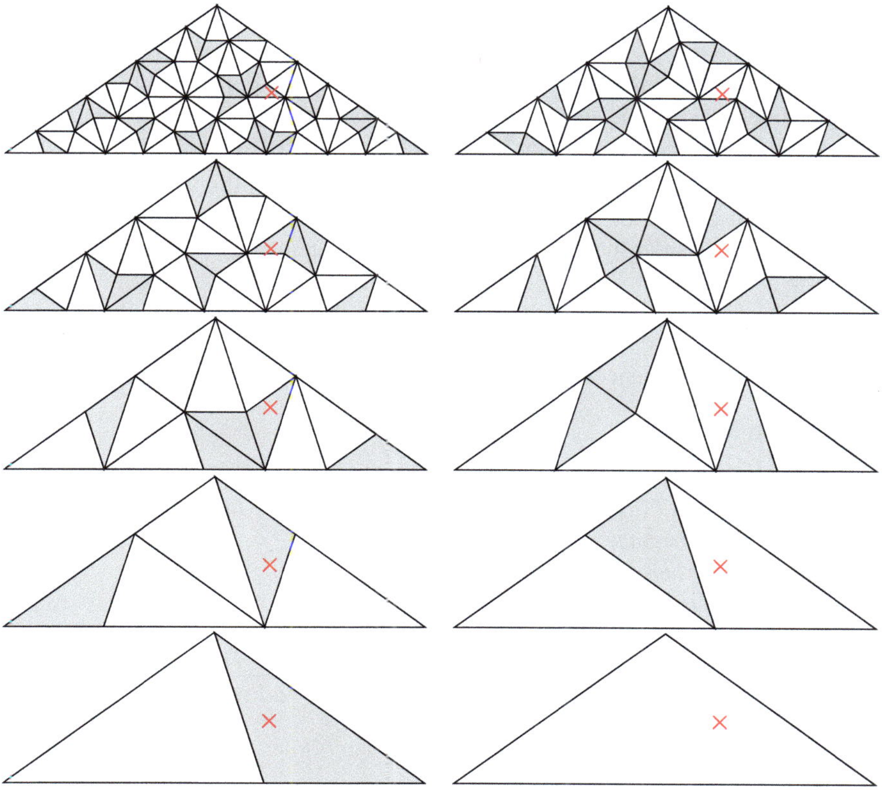

We see that the sequence starts with "10" repeated 5 times.

Finally, the Star tiling. This is obtained by applying $\mathcal{F}$ to the Sun, hence by the same reasoning in the proof of the Lemma 3.25, its index sequence is simply the one of the Sun shifted by 2 slots.

3.4.2 Classification of Tilings with Robinson Triangles

In this section, whenever we write $\iota(\mathcal{T}, z_0)$, we tacitly assume that $\mathcal{T}$ is a Robinson tiling and z_0 a point in the interior of a tile.

Lemma 3.26 *If $(a_n)_{n\in\mathbb{N}}$ is the index sequence of a pair $(\mathcal{T}, z_0)$ then, for all $k \geq 0$:*

$$a_k = 1 \quad \Longrightarrow \quad a_{k+1} = 0. \tag{3.19}$$

Proof Let $(T_0, T_1, T_2, \ldots)$ be the sequence of tiles containing z_0 defined at Sect. 3.4. For n even, if T_n is a golden gnomon, then after Composition 1 it will be inside a bigger golden gnomon. For n odd, if T_n is a golden triangle, then after Composition 2 it will be inside a bigger golden triangle. In both cases, $a_n = 1 \Longrightarrow a_{n+1} = 0$. ■

In the following, by an *angular sector* we mean the intersection of two closed half-planes whose boundary lines intersect at a point; the point of intersection will be called the *vertex* of the angular sector; by *angle* of an angular sector we mean the internal angle formed by the boundary lines of the two half-planes.

Recall that two tilings $\mathcal{T}$ and $\mathcal{T}'$ are equivalent, and in this case we will write $\mathcal{T} \sim \mathcal{T}'$, if one can be obtained from the other with a direct isometry (Definition 2.6). On the set $2^{\mathbb{N}}$, consider the following equivalence relation, called *tail equivalence*:

$$(a_n)_{n\in\mathbb{N}} \sim (b_n)_{n\in\mathbb{N}} \iff \exists\, n_0 \in \mathbb{N} : a_k = b_k \;\forall\, k \geq n_0.$$

Thus, two binary sequences are equivalent if and only if they are eventually equal.

Proposition 3.27

(i) Any $(a_n)_{n\in\mathbb{N}} \in 2^{\mathbb{N}}$ satisfying (3.19) *is the index sequence of a pair $(\mathcal{T}, z_0)$.*
(ii) $\iota(\mathcal{T}, z_0) \sim \iota(\mathcal{T}, z_0')$ for all z_0 and z_0' (the tail of the index sequence is independent of the base point).
(iii) If $\mathcal{T} \sim \mathcal{T}'$, then $\iota(\mathcal{T}, z_0) \sim \iota(\mathcal{T}', z_0')$.
(iv) The map ι induces a 2-to-1 function:

$$\frac{\{\textit{Robinson tilings of the plane}\}}{\textit{direct isometries}} \longrightarrow \frac{\{\textit{binary sequences satisfying (3.19)}\}}{\textit{tail equivalence}}.$$

Proof Let $(a_n)_{n\in\mathbb{N}} \in 2^{\mathbb{N}}$ be a sequence satisfying (3.19). First we convert the binary sequence into a sequence of tiles:

$$(T_0, T_1, T_2, \dots) \tag{3.20}$$

The shape of the triangles is determined by the sequence: if a_n and n have the same parity, then $[T_n]$ is a golden triangle, otherwise it is a golden gnomon.

Next, we fix the type and position. Choose arbitrarily the type and position of T_0. By induction, we shall show that this fixes the type and position of T_n for all n.

Assume, by inductive hypothesis, that this is true for some $n \geq 0$: thus, we know $T_n \subset \mathbb{C}$ (we know its shape, type, and position). Assume that n is even and $(a_n, a_{n+1}) = (0, 0)$. Thus T_{n+1} is a golden gnomon, and T_n is a golden triangle whose type and position we know. T_{n+1} is the Composition 1 of a large and a small triangle:

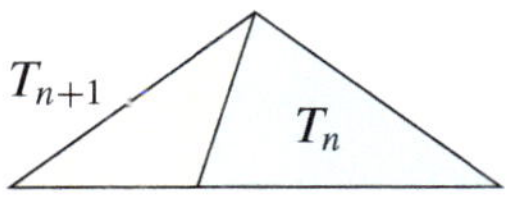

and there is only one way to place T_{n+1} in the plane so that its large sub-triangle coincides with T_n. The position of T_{n+1} is then fixed. It follows from the composition rule that if T_n is of type A (like in the picture) then T_{n+1} must be of type B, and if T_n is of type B then T_{n+1} must be of type A. The type of T_{n+1} is then fixed as well.

One can repeat the same argument for all possible values of (a_n, a_{n+1}), for both even and odd n, and check in all possible cases that the type and position of T_{n+1} are uniquely determined by the type and position of T_n.

From the sequence (3.20) we now construct a tiling of the plane $\mathcal{T}$. Call $\mathcal{T}_n$ the partial tiling obtained by applying n times Decomposition to T_{2n}. Then, $\mathcal{T}_n \subset \mathcal{T}_{n+1}$ for all n and the union

$$\bigcup_{n \geq 0} \mathcal{T}_n \tag{3.21}$$

is a tiling covering exactly the set

$$X := \bigcup_{n \geq 0} T_{2n}. \tag{3.22}$$

We now show that X is either the whole plane, a half-plane or an angular sector with angle $\pi/5$, and that in the latter two cases there exists a <u>unique way</u> to extend (3.21) to a tiling of the plane, thus in particular proving the claim (i).

Before that, call z_0 the barycenter of T_0 and observe that, if a tiling $\mathcal{T}$ of the plane extending (3.21) exists, then by construction $\iota(\mathcal{T}, z_0) = (a_n)_{n \in \mathbb{N}}$ is the binary sequence we started from. Observe also that:

(a) The partial tiling (3.21) is uniquely determined up to a direct isometry (the position of T_0, that we can choose arbitrarily) and a reflection (the type of T_0, which is not determined by the index sequence). In fact, the only important thing is the <u>tail</u> of the index sequence: to reconstruct arbitrarily large portions of the tiling (3.21) it is enough to know T_n for arbitrarily large n.

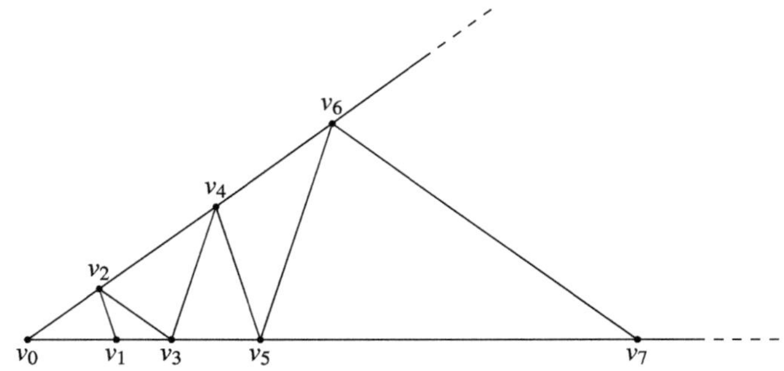

Fig. 3.10 Proof of Proposition 3.27. Case (1)

(b) Given any two points z_0 and z_0' in the set (3.22), there is n large enough such that the two points are contained in the same triangle T_{2n}. In particular, this means that the two points give two index sequences that are eventually equal.

From (a), and from the uniqueness of the extension of (3.21) to a tiling of the plane, it will follow that: (iii) the equivalence class of an index sequence depends only on the equivalence class of a tiling, and (iv) there are exactly two equivalence classes of tilings for each class of index sequences (the map at point (iv) would be 1-to-1 if we were quotienting by isometries rather than only direct isometries). From (b) it will follow that any two points in the set X give rise to equivalent index sequences, which is almost the claim (ii).

We distinguish three cases:

(1) there exists $n_0 \in \mathbb{N}$ such that all triangles T_n with $n \geq n_0$ share a common vertex;
(2) there exists $n_0 \in \mathbb{N}$ and a segment s, such $s \subset T_n \setminus \mathring{T}_n$ for all $n \geq n_0$;
(3) none of the above.

Case (1). Call v_0 the common vertex and consider the angular sector delimited by the half-lines from v_0 extending the edges of T_{n_0}. An example is in Fig. 3.10. In the picture, the first tile T_{n_0} is an L_B tile with vertices v_0, v_1, v_2; the next tile is an S_A tile with vertices v_0, v_2, v_3; the next is a bigger S_A tile with vertices v_0, v_3, v_4; etc.

In general, for $n \geq n_0$, the (internal) angle of T_n at v_0 cannot be $2\pi/5$ or $3\pi/5$. Indeed, T_n is obtained from T_{n+2k} by Decomposition (for all $k \geq 1$), and any internal angle of a tile after enough Decompositions splits into a number of $\pi/5$ angles (the $3\pi/5$ angle of an S tiles splits into $2\pi/5 + \pi/5$, and any $2\pi/5$ angle of an L tile splits into $\pi/5 + \pi/5$ after at most two Decompositions). We conclude that the angular sector forms an angle of $\pi/5$ and all T_n, for $n \geq n_0$, are contained in such an angular sector. We must prove that they fill the sector, but this is obvious since we have arbitrarily large triangles wedged between the two half-line.

The partial tiling can be completed to a tiling of the plane as follows. Call ℓ_1 and ℓ_2 the half-lines delimiting the angular sector. After reflection, we get another tiling of an angular sector, as illustrated in the next picture:

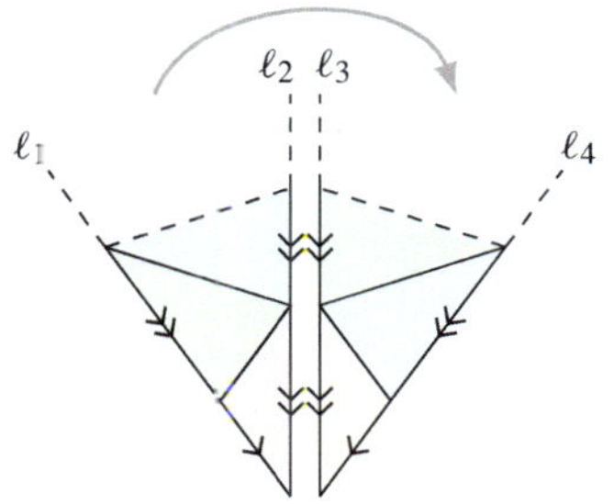

Now, the edges on the half-line ℓ_2 match the edges on the half-line ℓ_3, and we can glue these two tilings to form a tiling of an angular sector with angle $2\pi/5$. Since the edges on the half-line ℓ_1 match the edges on the half-line ℓ_4, we can now take five copies of this partial tiling and glue them together to form a tiling of the plane, that will be either a Sun or a Star depending on the sequence (3.20).

This proves that there exists a tiling of the plane extending (3.21). We now show that this tiling is unique.

Let $\mathcal{T}$ be any tiling of the plane extending (3.21) and call $\mathcal{N}_n$ the neighborhood of z_0 in the tiling obtained from $\mathcal{T}$ by applying Composition n times. Since $\mathcal{N}_n$ is obtained from $\mathcal{N}_{n+3}$ applying Decomposition 3 times, it follows from (3.12) that it must be either a Sun or a Star, for every $n \geq n_0$. From these neighborhoods one gets patches with 5-fold symmetry covering the plane, proving that $\mathcal{T}$ is either a Sun or a Star tiling, depending on T_{n_0} (Proposition 3.22).

Case (2). An instance of the second case is in Fig. 3.11. In the picture, the first tile T_{n_0} is an L_B tile with vertices v_0, v_1, v_2; the next tile is an S_A tile with vertices v_0, v_2, v_3; the next is an inflated L_A tile with vertices v_0, v_3, v_4; then we have an inflated S_B tile with vertices v_3, v_4, v_5; etc.

In general, call s the edge of T_{n_0} which is contained in the boundary of all the subsequent triangles and call v_0 and v_1 the two extreme points. Up to a direct isometry, we can assume that $v_0 = 0$ and $v_1 > 0$. For all $n \geq n_0$, by hypothesis: (a) T_n has one edge on the real line, call $v_{0,n}, v_{1,n} \in \mathbb{R}$ the extreme points of this edge, with $v_{0,n} < v_{1,n}$; (b) T_n must be contained in the same half-plane of T_{n_0} (since the triangles are one inside the other). We can assume that they are in the upper half-plane, like in the picture.

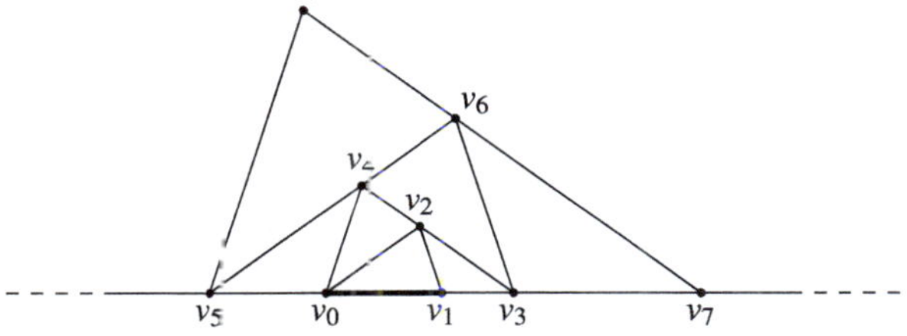

Fig. 3.11 Proof of Proposition 3.27. Case (2)

We now prove that either all triangles T_m, for a suitable $m_0 \geq n_0$ and all $m \geq m_0$, have a common vertex, and we are back to case (1), or the union (3.22) contains the real line. It follows from the subdivision rule that, for every n, either $v_{1,n+1} = v_{1,n}$ or $v_{1,n+1} \geq v_{1,n} + 1$. Thus, the monotonic sequence $(v_{1,n})_{n \geq n_0}$ is either divergent (and (3.22) contains the half-line from v_0 to $+\infty$) or eventually stabilizes: that is, $v_{1,m} = v_{1,m_0}$ for some $m_0 \geq n_0$ and all $m \geq m_0$. In the latter case, v_{1,m_0} is a common vertex to all triangles T_m with $m \geq m_0$. Similarly, $(v_{0,n})_{n \geq n_0}$ is either divergent, which means that (3.22) contains the half-line from v_0 to $-\infty$), or eventually stabilizes, which means that we are back to case (1).

Next, under the assumption that the triangles T_n cover the real line, we prove that they cover the whole upper half-plane. Let T^x be a golden gnomon with base on the real line and vertices $\pm\varphi x$ and $(0, x\sqrt{3-\varphi})$. For every x, there exists n big enough so that T_{2n} contains the points $\pm\varphi x$ on the real line; now, whatever is the shape of T_{2n}, if it contains the points $\pm\varphi x$, it also contains the triangle T^x. Thus,

$$\bigcup_{n\in\mathbb{N}} T_{2n} \supseteq \bigcup_{x>0} T^x,$$

but the latter union is clearly the whole upper half-plane.

Similarly to case (1), one can complete (3.21) to a tiling of the whole plane by reflecting it across the boundary line. We now show that this is the only way to complete (3.21) to a tiling of the whole plane.

In the notations above, let y_0 be any internal point to the segment s: this is contained in an edge of T_n for all $n \geq n_0$.

Let $\mathcal{T}$ be any tiling of the plane extending (3.21) and call $\mathcal{N}_n$ the neighborhood of y_0 in the tiling obtained from $\mathcal{T}$ by applying Composition n times. Since y_0 is internal to an edge, each neighborhood will be formed by two adjacent tiles (bigger and bigger). The first five admissible patches in (3.2a)–(3.2b) are invariant under reflection across the common edge. The trapezoids in (3.2b) after Decomposition will contain a vertex neighborhood that is invariant under reflection:

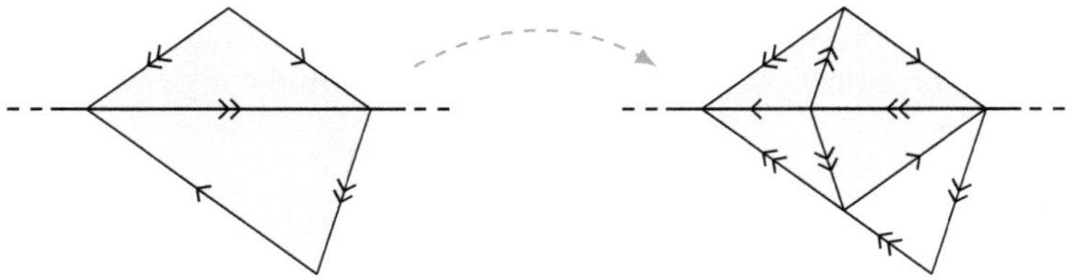

Finally, the triangles in (3.2b) will contain a trapezoid after Decomposition:

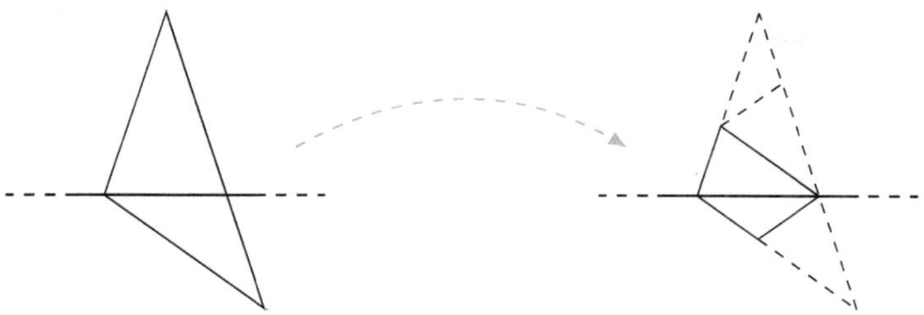

and then a neighborhood invariant under reflection after another Decomposition. Thus, after at most two Decompositions, we will get a neighborhood of y_0 that is invariant under reflection.

Since $\mathcal{N}_n$ is obtained from $\mathcal{N}_{n+2}$ applying Decomposition twice, it follows from the observation above that it must be invariant under reflection across the real axis, for every $n \geq n_0$. From these neighborhoods one gets nested patches with reflection symmetry covering the plane, proving that $\mathcal{T}$ is obtained from the partial tiling (3.21) by reflection.

Case (3). An instance of the third case is in Fig. 3.12. In the picture, the first tile T_{n_0} is an L_B tile with vertices v_0, v_1, v_2; the next tile is an inflated L_B tile with vertices v_0, v_1, v_3; the next is an inflated S_A tile with vertices v_1, v_3, v_4; then we have an inflated L_A tile with vertices v_3, v_4, v_5; etc.

In general, observe that by assumption, for every n and every boundary point x of T_n there exists a k large enough such that x is not on the boundary of T_{n+k}. This means that every point of T_n is internal in T_{n+k}, and then every point of the set X in (3.22) is internal in X. Thus, X is open in $\mathbb{C}$. On the other hand, it follows from Lemma 2.20(ii) that X is also closed in $\mathbb{C}$.[1] Since X is both open and close in $\mathbb{C}$, and $\mathbb{C}$ is connected, we deduce that $X = \mathbb{C}$.

At this point, the proof of the proposition is almost complete. About point (ii), we proved that $\iota(\mathcal{T}, z_0) \sim \iota(\mathcal{T}, z_0')$ for all z_0 and z_0' in the set X in (3.22) (which is the same for the two index sequences, since it depends only on their tail). If we are in case (3), the proof of (ii) is completed.

In case (2), if z_0 belongs to X and z_0' to $\mathbb{C} \smallsetminus X$, call z_0'' the point obtained from z_0' by a reflection σ across the boundary line of X. Clearly $\iota(\mathcal{T}, z_0') = \iota(\sigma\mathcal{T}, \sigma(z_0')) = \iota(\sigma\mathcal{T}, z_0'')$. But by construction $\sigma\mathcal{T} = \mathcal{T}$, hence $\iota(\mathcal{T}, z_0') = \iota(\mathcal{T}, z_0'') \sim \iota(\mathcal{T}, z_0)$, where the last equivalence follows from the fact that now both z_0 and z_0'' belong to X.

In case (1), if $z_0 \in X$ and $z_0' \in \mathbb{C} \smallsetminus X$, either z_0' belongs to a reflection of X, or to a patch obtained from X with a rotation of $2\pi/5$. Then, one proceeds like in case (2) and uses the fact that $\mathcal{T}$ is invariant under both reflections (across any of the boundary half-lines of X) and $2\pi/5$ rotations (with respect to the vertex of X). ■

[1] Observe that the nested sequence (3.20) is not locally finite. But X is also a union $X = \bigcup_\alpha t_\alpha$ of tiles t_α of the partial tiling (3.21), and this is a locally finite collection of closed sets.

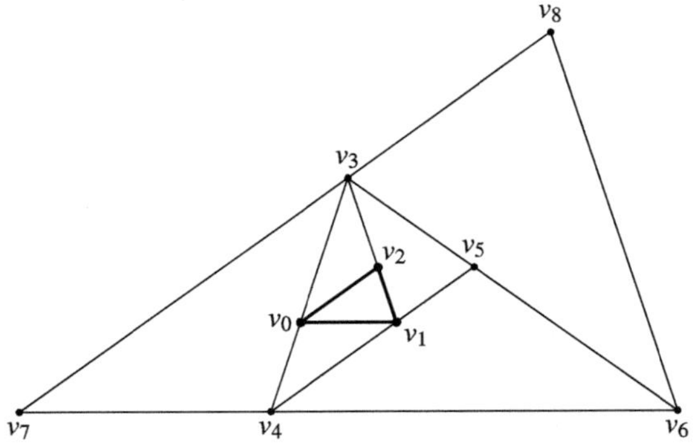

Fig. 3.12 Proof of Proposition 3.27. Case (3)

From point (a) in the proof of Proposition 3.27 we see that two different equivalence classes of tilings that correspond to the same class of index sequences must be related by a reflection. We can then rephrase Proposition 3.27(iv) as follows.

Proposition 3.28 *The map ι induces a bijection:*

$$\frac{\{\textit{Robinson tilings of the plane}\}}{\textit{isometries}} \xrightarrow{1:1} \frac{\{\textit{binary sequences satisfying (3.19)}\}}{\textit{tail equivalence}}.$$

An immediate consequence of Proposition 3.27(iv) is that there are uncountably many inequivalent Robinson tilings (see Proposition 3.29 below). In this set of cardinality $|\mathbb{R}|$, there are exactly two classes of tilings with 5-fold symmetry (Proposition 3.22). On the other hand, there are infinitely many inequivalent tilings with reflection symmetry (cf. Proposition 3.31 below).

Proposition 3.29 *The set of equivalence classes of Robinson tilings has the cardinality of the continuum.*

Proof Let $\mathbb{S}$ be the set of binary sequences satisfying (3.19). We shall now compute the cardinality of the set $\mathbb{S}/\!\sim$.

The map $2^{\mathbb{N}} \to \mathbb{S}$ replacing every 1 in a binary sequence by a pair 10 is a bijection. The inverse map, removing from the sequence all the 0's that follow a 1, is well-defined by definition of $\mathbb{S}$. Thus, $|\mathbb{S}| = |2^{\mathbb{N}}| = |\mathbb{R}|$, and obviously $|\mathbb{S}/\!\sim| \leq |\mathbb{S}|$.

For the opposite inequality, it is enough to show that the fibers of the quotient map $\mathbb{S} \to \mathbb{S}/\!\sim$ are countable sets (this is a consequence, e.g., of [HSW10, Theorem 1.5.14](c)). If $(a_n)_{n\in\mathbb{N}} \in \mathbb{S}$, its equivalence class is given by

$$\bigcup_{n_0\in\mathbb{N}} \big\{(x_n)_{n\in\mathbb{N}} \in \mathbb{S} : x_k = a_k \ \forall\, k \geq n_0\big\},$$

and this is a countable union of finite sets. ■

Let us call *skeleton* of a Robinson tiling the union of the edges of all its tiles.

Proposition 3.30 *In a Robinson tiling, if the skeleton contains a full line, then this line must be an axis of symmetry (the tiling must be invariant under reflection across this line).*

Proof Let $\mathcal{T}$ be the tiling and ℓ the line in the statement of the proposition. Choose one of the two half-planes delimited by ℓ. The subset of tiles belonging to this half-plane is a partial tiling covering exactly the half-plane, and we saw in the proof of Proposition 3.27, case (2), that the only way to complete it to a tiling of the plane is by reflection across ℓ, proving that ℓ is an axis of symmetry for $\mathcal{T}$. ■

Proposition 3.31 *(i) Not every Robinson tiling has reflection symmetry. (ii) The set of equivalence classes of Robinson tilings with reflection symmetry has the cardinality of the continuum.*

Proof (i) If a tiling has reflection symmetry, any nested sequence of tiles (3.20) is contained in a half-plane. Any instance of Case (3) in the proof of Proposition 3.27 has no reflection symmetry. Any tiling obtained with iterated inflation and Decomposition from a single tile belongs to case (3), for example the golden triangle and the golden gnomon in Sect. 3.1.3.

(ii) The substitution

$$0 \mapsto \underbrace{00000000}_{\alpha}\,, \qquad 1 \mapsto \underbrace{00100010}_{\beta}\,,$$

gives a bijection between arbitrary binary sequences and sequences made of blocks α and β. The set of classes of $\{\alpha, \beta\}$-sequences, modulo tail equivalence, has then cardinality $|\mathbb{R}|$. We now show that each sequence of this form is the index sequence of a tiling which is symmetric under reflection.

The index sequence tells us how to reconstruct (part of) the tiling. The first tile should be either L_A or L_B ($a_0 = 0$). Assume it is L_A (the other case being similar). Up to a direct isometry, we can assume that the triangle has its right leg on the real axis:

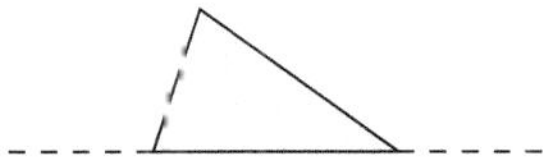

By inductive hypothesis, assume that after $8n$ Compositions, T_{8n} is similar to the tile above (an L_A tile with right leg on the real axis). We are going to prove that $T_{8(n+1)}$ is also similar to an L_A tile with right leg on the real axis, so that the union over n is entirely contained in an half-plane.

We have to consider two cases: either $(a_{8n}, \ldots, a_{8n+7})$ is a block α or a block β (observe that $a_{8n+8} = 0$ in any case, since both blocks start with zero). Thus, in the first case:

$$(a_{8n+1}, \ldots, a_{8n+8}) = (0, \ldots, 0).$$

By inductive hypothesis we start with an (inflated) L_A tile placed as explained above and apply this sequence of instructions to extend the patch. After two steps we get:

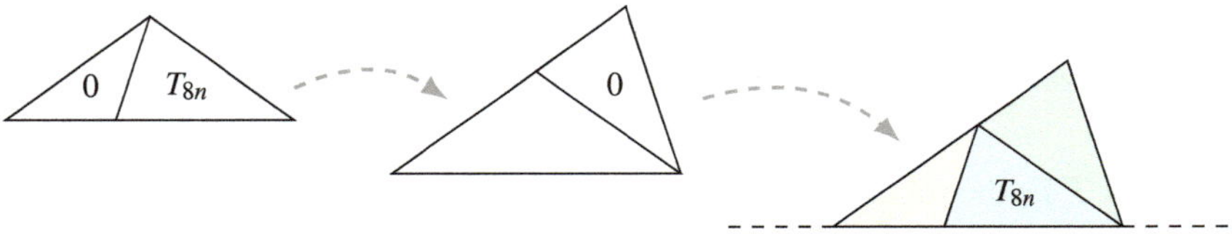

so that T_{8n+2} is an inflated L_B tile with left leg on the real axis. Similarly, T_{8n+4} is an inflated L_A tile with right leg on the real axis. After another four steps we get an inflated L_A tile with right leg on the real axis.

We pass to the second case. Now:

$$(a_{8n+1}, \ldots, a_{8n+8}) = (0, 1, 0, 0, 0, 1, 0, 0).$$

After the first four steps we get:

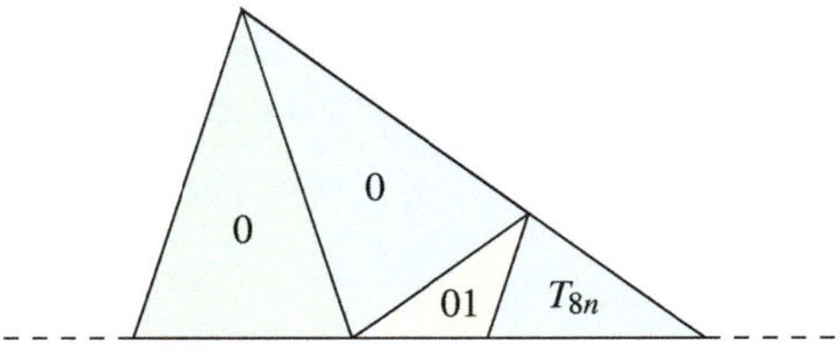

Thus, T_{8n+4} is an inflated L_B tile with right leg on the real axis. Similarly, after four more steps, applying the instruction 0100 one more time we get an inflated L_A tile with right leg on the real axis. ■

3.5 On the Density of Large and Small Triangles

We want to show that, in a Robinson tiling, the density of S and L triangles is proportional to their area. This may seem a bit counterintuitive, since it means that the tile that occupies more space appears more often. Let us introduce some notations and make this statement precise.

Suppose we are given a Robinson tiling $\mathcal{T}$. Given a patch $\mathcal{P}$, we denote by $s(\mathcal{P})$ the number of S triangles and by $\ell(\mathcal{P})$ the number of L triangles in $\mathcal{P}$. Similarly, given a compact subset K of $\mathbb{R}^2$, we denote by $s(K)$ the number of S triangles and by $\ell(K)$ the number of L triangles that are contained in K.

Let $\mathcal{T}'$ be the tiling obtained from $\mathcal{T}$ by applying $k \geq 0$ Compositions. We will say that a patch $\mathcal{P} \subset \mathcal{T}$ has a *k-hierarchical structure* if there is a patch $\mathcal{P}' \subset \mathcal{T}'$ such that

$$\bigcup_{T \in \mathcal{P}} T = \bigcup_{T' \in \mathcal{P}'} T'.$$

The other way round, this means that $\mathcal{P}$ is a obtained from $\mathcal{P}'$ by applying Decomposition k times.

The situation is illustrated in Fig. 3.13, where we in the background see a portion of a Robinson tiling and the thick lines in the foreground indicate the result of Composition applied k times ($k = 2$ in the picture). A collection of small L and S triangles has a k-hierarchical structure if the region covered by such triangles is also exactly covered by big $\varphi^k L$ and $\varphi^k S$ triangles (like the red area in Fig. 3.13).

Proposition 3.32 *Let $\mathcal{T}$ be a Robinson tiling and $\mathcal{E} \subset \mathcal{T}$ a patch with a k-hierarchical structure, with $k \geq 2$. Then*

$$\frac{F_{2k}}{F_{2k-1}} \leq \frac{\ell(\mathcal{E})}{s(\mathcal{E})} \leq \frac{F_{2k+1}}{F_{2k}} \tag{3.23}$$

where F_k is the kth Fibonacci number (cf. Appendix A.2). Moreover, for $k \to \infty$:

$$\ell(\mathcal{E}) = \frac{4}{\sqrt{5\varphi + 10}} \operatorname{Area}(R)\varphi^k + O(\varphi^{-2k}), \tag{3.24a}$$

$$s(\mathcal{E}) = \frac{4}{\sqrt{5\varphi + 10}} \operatorname{Area}(R)\varphi^{k-1} + O(\varphi^{-2k}), \tag{3.24b}$$

where R is the union of tiles of $\mathcal{E}$.

Observe that the upper and lower bound in (3.23) converge very fast to φ. From Cassini's identity (A.15), it follows that their difference is

$$\frac{F_{2k+1}}{F_{2k}} - \frac{F_{2k}}{F_{2k-1}} = \frac{1}{F_{2k} F_{2k-1}} = O(\varphi^{-4k}).$$

For example, if $k = 3$ the error in replacing the bounds by φ is already less than 1%.

Proof By hypothesis, $\mathcal{E}$ is a finite collection of small tiles whose union R is also a union of big tiles. Let us count the number of L and S tiles in $\mathcal{E}$.

Suppose that the region R is a union of n_1 tiles $\varphi^k L$ and n_2 tiles $\varphi^k S$. With a slight abuse of notations

$$\frac{\ell(\mathcal{E})}{s(\mathcal{E})} = \frac{n_1\, \ell(\varphi^k L) + n_2\, \ell(\varphi^k S)}{n_1\, s(\varphi^k L) + n_2\, s(\varphi^k S)} = \lambda \frac{\ell(\varphi^k L)}{s(\varphi^k L)} + (1 - \lambda) \frac{\ell(\varphi^k S)}{s(\varphi^k S)} \tag{3.25}$$

where

$$\lambda := \frac{n_1 s(\varphi^k L)}{n_1 s(\varphi^k L) + n_2 s(\varphi^k S)}.$$

Since $0 \leq \lambda \leq 1$, from (3.25) we get

$$\min\left\{\frac{\ell(\varphi^k L)}{s(\varphi^k L)}, \frac{\ell(\varphi^k S)}{s(\varphi^k S)}\right\} \leq \frac{\ell(\mathcal{E})}{s(\mathcal{E})} \leq \max\left\{\frac{\ell(\varphi^k L)}{s(\varphi^k L)}, \frac{\ell(\varphi^k S)}{s(\varphi^k S)}\right\}. \tag{3.26}$$

We now compute the number of small tiles in each big tile.

Let us start with $\varphi^k L$. Let a_m be the number of L triangles and b_m the number of S triangles in the patch obtained from $\varphi^k L$ by applying Decomposition m times, with $0 \le m \le k$. Thus, $a_0 = 1, b_0 = 0$, and since after Decomposition each large triangle gives two large triangles and one small, and each small triangle gives one large and one small, we have the recursive relations:

$$a_{m+1} = 2a_m + b_m, \qquad b_{m+1} = a_m + b_m,$$

for all $0 \le m < k$. From the recursive relation for Fibonacci numbers (see Appendix A.2), we see that $a_m = F_{2m+1}$ and $b_m = F_{2m}$ solve the recurrence problem above. In particular,

$$\frac{\ell(\varphi^k L)}{s(\varphi^k L)} = \frac{a_k}{b_k} = \frac{F_{2k+1}}{F_{2k}}. \tag{3.27}$$

The computation for $\varphi^k S$ is similar: we have the same recursive relations, but we start with $a_0 = 0$ and $b_0 = 1$. We find $a_m = F_{2m}$ and $b_m = F_{2m-1}$ for all $1 \le m \le k$, and in particular

$$\frac{\ell(\varphi^k S)}{s(\varphi^k S)} = \frac{a_k}{b_k} = \frac{F_{2k}}{F_{2k-1}}. \tag{3.28}$$

If we now insert (3.27) and (3.28) in (3.26), and use (A.16) to establish which fraction is bigger, we arrive at the inequalities (3.23).

Next, in the notations above, observe that the area of R is n_1 times the area of $\varphi^k L$ plus n_2 times the area of $\varphi^k S$ triangle. From (A.7):

$$\text{Area}(R) = (n_1\varphi + n_2)\,\varphi^k\,\frac{1}{4}\sqrt{\varphi + 2}.$$

On the other hand, from (A.12):

$$\ell(\mathcal{E}) = n_1 F_{2k+1} + n_2 F_{2k} = (n_1\varphi + n_2)\frac{\varphi^{2k}}{\sqrt{5}} + O(\varphi^{-2k}),$$

$$s(\mathcal{E}) = n_1 F_{2k} + n_2 F_{2k-1} = (n_1\varphi + n_2)\frac{\varphi^{2k-1}}{\sqrt{5}} + O(\varphi^{-2k}),$$

hence the formulas (3.24). ■

We now pass to more general subsets of $\mathbb{R}^2$. Let $K \subset \mathbb{R}^2$ be a compact convex set, $K \neq \varnothing$. Recall that the *perimeter* of K can be defined as the supremum of the perimeters of all the inscribed convex polygons; similarly, the *area* of K can be defined as the supremum of the areas of all the inscribed convex polygons [Lay82].

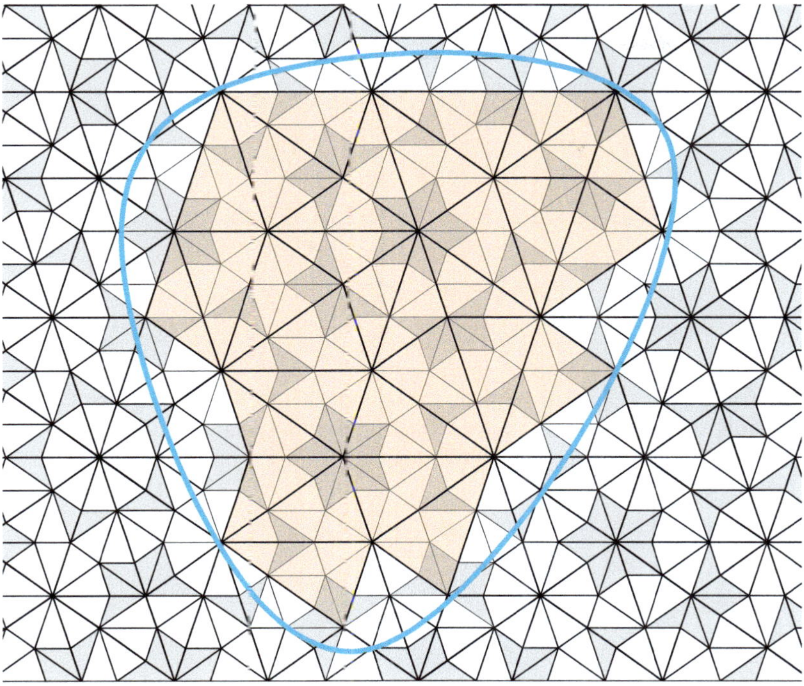

Fig. 3.13 Proof of Lemma 3.33

Lemma 3.33 *Let $K \subset \mathbb{R}^2$ be a non-empty compact convex set with perimeter P, $\mathcal{T}$ a Robinson tiling, $k \geq 0$, $\mathcal{E}$ the collection of all $\varphi^k L$ and $\varphi^k S$ tiles inside K (if any) and $R \subseteq K$ their union. Then,*

$$\text{Area}(K \setminus R) \leq \varphi^k P. \tag{3.29}$$

Proof An illustration is in Fig. 3.13. In blue we see the boundary curve of the set K, inside in red the region R, union of all big tiles that lie completely inside K.

Since K can be approximated with arbitrary precision by a convex polygon, it is enough to do the proof when it is a polygon. Every convex polygon is an intersection of half-planes. Let E be the set of edges of K. For each edge $e \in E$ consider the half-plane Π_e bounded by the line through e and containing the polygon K:

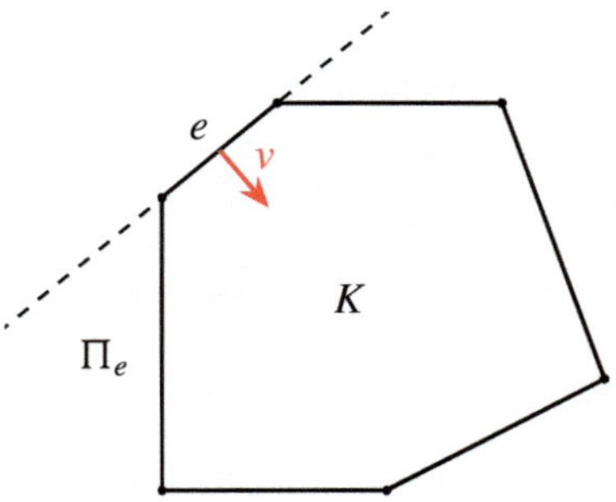

One has $\bigcap_{e \in E} \Pi_e = K$. Now we push the boundary line of each half-plane Π_e inside the polygon, with a translation in the direction v orthogonal to its boundary and of a length φ^{k+1}. We call Π'_e the new half-plane:

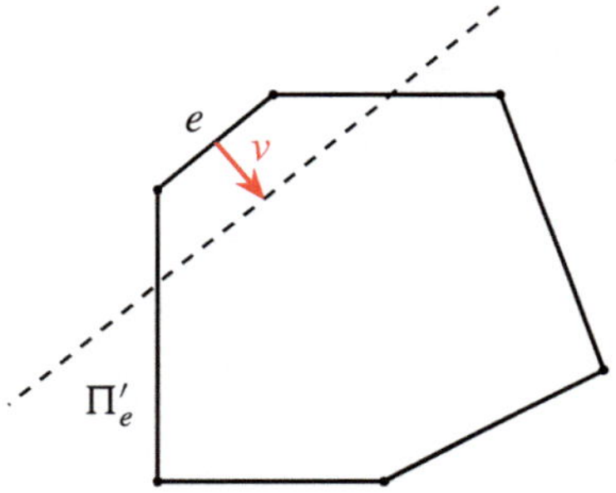

Points inside Π'_e are at distance at least φ^{k+1} from the edge e. The intersection $\bigcap_{e \in E} \Pi'_e$ is a new convex polygon K':

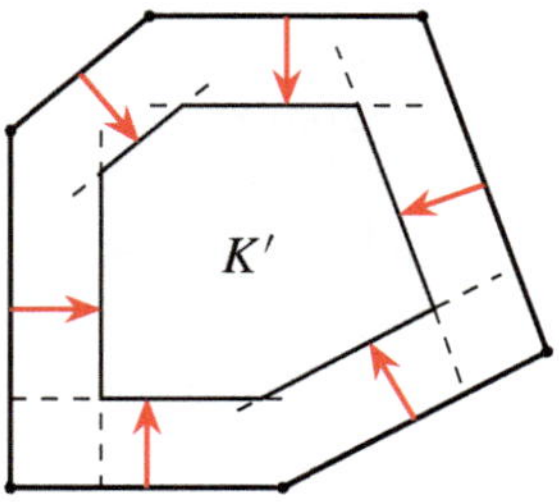

If a $\varphi^k L$ or a $\varphi^k S$ tile intersects K', since the longest edge has length φ^{k+1} and the boundary points of K are at distance $\geq \varphi^{k+1}$ from K', it follows that all three vertices of the tile lie inside K, and then the whole tile lie inside K (since K is convex). Thus, every big tiles intersecting K' belongs to $\mathcal{E}$, and we have

$$K' \subseteq R \subseteq K.$$

In particular, Area$(K \setminus R) \leq$ Area$(K \setminus K')$. We now compute an upper bound for the latter area. (We stress that the Lemma holds, and the proof works, even when R and/or K' are the empty set.)

For each $e \in E$ consider the rectangle H_e built on that edge and with height φ^{k+1}:

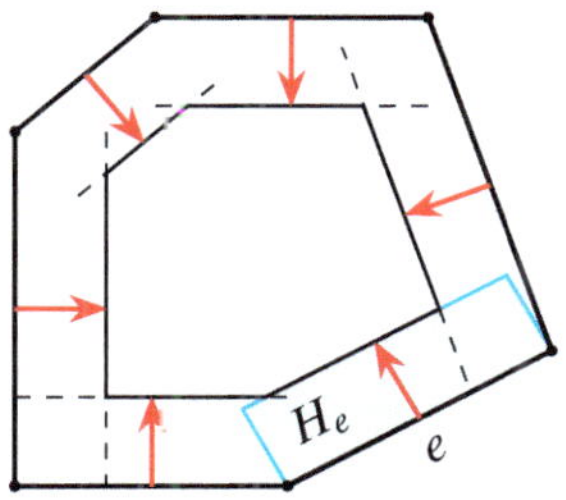

Since K is convex, the internal angles are no greater than π, and the union $\bigcup_{e \in E} H_e$ covers the region $K \setminus K'$. Thus

$$\text{Area}(K \setminus K') \leq \sum_{e \in E} \text{Area}(H_e) = \sum_{e \in E} |e| \cdot \varphi^{k+1} = \varphi^{k+1} P,$$

where P is the perimeter of K. ■

Proposition 3.34 *Let $(K_n)_{n \in \mathbb{N}}$ be a sequence of non-empty compact convex subsets of $\mathbb{R}^2$. Denote by A_n the area and P_n the perimeter of K_n, and assume that*

$$\lim_{n \to \infty} \frac{A_n}{P_n} = \infty. \tag{3.30}$$

Then, for any Robinson tiling:

$$\lim_{n \to \infty} \frac{\ell(K_n)}{s(K_n)} = \varphi. \tag{3.31}$$

In the terminology of [GS87], we express the independence of the limit (3.31) of the sequence by saying that Robinson tilings are *metrically prototile balanced.*

Note that $A_n/P_n \leq P_n/4\pi$ (by the isoperimetric inequality, see [Lay82, Chap. 13]). Thus, (3.30) implies that both $A_n \to \infty$ and $P_n \to \infty$, and the regions that we are considering are bigger and bigger.

Proof Let

$$k(n) := \frac{1}{2} \sup \left\{ j \in \mathbb{N} : A_n/P_n > \varphi^j \right\}.$$

Due to (3.30), $k(n)$ is well-defined for n big enough (the set over which we take the sup is non-empty). Moreover, $k(n) \to \infty$ for $n \to \infty$.

Let R_n be the union of all $\varphi^{k(n)}L$ and $\varphi^{k(n)}S$ tiles inside K_n. The decomposition $K_n = R_n \cup (K_n \setminus R_n)$ gives:

$$\frac{\ell(K_n)}{s(K_n)} = \frac{\ell(R_n)}{s(R_n)} \cdot \frac{1 + \dfrac{\ell(K_n \setminus R_n)}{\ell(R_n)}}{1 + \dfrac{s(K_n \setminus R_n)}{s(R_n)}}.$$

On one side, from (3.23) it follows that $\lim_{n\to\infty} \ell(R_n)/s(R_n) = \varphi$. Now, we prove that the numerator and denominator of the remaining fraction tend to 1. From (3.29):

$$\ell(K_n \setminus R_n) \le \frac{\text{Area}(K_n \setminus R_n)}{\text{Area}(L)} \le \frac{4\varphi^{k(n)} P_n}{\varphi\sqrt{\varphi + 2}},$$
$$s(K_n \setminus R_n) \le \frac{\text{Area}(K_n \setminus R_n)}{\text{Area}(S)} \le \frac{4\varphi^{k(n)} P_n}{\sqrt{\varphi + 2}}.$$

From (3.24):

$$\ell(R_n) = c^{-1}\,\text{Area}(R_n)\varphi^{k(n)} + O(\varphi^{-2k(n)}),$$
$$s(R_n) = c^{-1}\,\text{Area}(R_n)\varphi^{k(n)-1} + O(\varphi^{-2k(n)}),$$

where $c := \frac{1}{4}\sqrt{5\varphi + 10}$ is a constant. It is then enough to show that

$$\frac{\varphi^{k(n)} P_n}{\text{Area}(R_n)\varphi^{k(n)}} \to 0. \tag{3.32}$$

By definition of $k(n)$ and (3.29) one has:

$$\text{Area}(R_n) = A_n - \text{Area}(K_n \setminus R_n) \ge A_n - \varphi^{k(n)} P_n \ge (\varphi^{2k(n)} - \varphi^{k(n)}) P_n.$$

Therefore

$$\frac{P_n}{\text{Area}(R_n)} \le \frac{1}{\varphi^{2k(n)} - \varphi^{k(n)}},$$

and the right hand side goes to 0 when $k(n) \to \infty$, thus proving (3.32). ∎

The next two lemmas will be useful in Chap. 4.

Lemma 3.35 *Let $(K_n)_{n\in\mathbb{N}}$ be a sequence of non-empty compact convex subsets of $\mathbb{R}^2$ such that $\lim_{n\to\infty} A_n/P_n = \infty$, where A_n is the area and P_n is the perimeter of K_n. Then, for any Robinson tiling:*

$$\lim_{n\to\infty} \frac{A_n}{\ell(K_n)} = c, \qquad \lim_{n\to\infty} \frac{A_n}{s(K_n)} = c\,\varphi, \tag{3.33}$$

where $c := \frac{1}{4}\sqrt{5\varphi + 10}$ is a constant.

Proof Let R_n be the union of all tiles inside K_n. On one side

$$\text{Area}(R_n) = \big(\varphi\,\ell(K_n) + s(K_n)\big)\text{Area}(S) = c\,\frac{\varphi\,\ell(K_n) + s(K_n)}{\varphi + \varphi^{-1}}.$$

From Proposition 3.34 it follows that

$$\lim_{n\to\infty} \frac{\text{Area}(R_n)}{\ell(K_n)} = c.$$

On the other hand, from (3.29), for $k = 1$, we deduce that

$$0 \le 1 - \frac{\text{Area}(R_n)}{A_n} = \frac{\text{Area}(K_n \setminus R_n)}{A_n} \le \varphi \frac{P_n}{A_n} \to 0$$

This gives the first limit in (3.33):

$$\lim_{n\to\infty} \frac{A_n}{\ell(K_n)} = \lim_{n\to\infty} \frac{\text{Area}(R_n)}{\ell(K_n)} \frac{A_n}{\text{Area}(R_n)} \to c \cdot 1.$$

The second limit in (3.33) is proved similarly. ■

Lemma 3.36 *Given a Robinson tiling and a non-empty compact convex set $K \subset \mathbb{R}^2$, denote by $\ell'(K)$ (resp. $s'(K)$) the number of L triangles (resp. S triangles) that intersect the boundary of K. Then*

$$\ell'(K) \le c'(P + \pi) \quad \text{and} \quad s'(K) \le c'\varphi\,(P + \pi), \tag{3.34}$$

where $c' := 8/\sqrt{\varphi + 2}$ is a constant.

Proof For the proof we modify the construction in Lemma 3.33 as follows. As before, we can assume that K is a convex polygon. We construct a set $K'' \supset K$ by translating each edge of K by φ in the outward direction orthogonal to the edge, and drawing an arc, centered at each vertex of K and with radius φ, to fill the blanks:

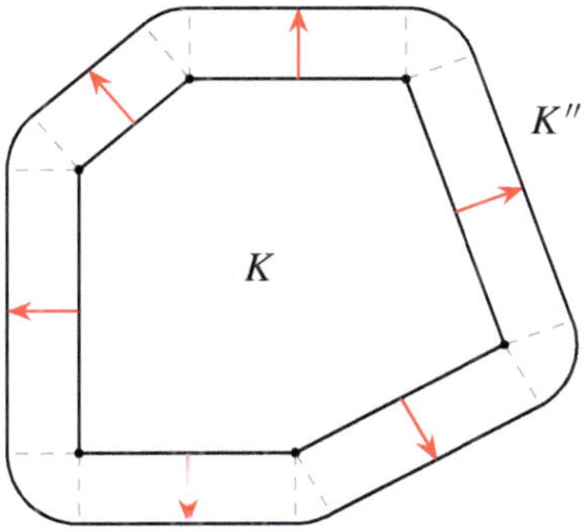

(3.35)

Points on the boundary of K'' are at distance no less than φ from K. Every L or S triangle intersecting the boundary of K is completely contained in $K'' \setminus K'$, where K' is the (smaller) convex polygon in the proof of Lemma 3.33. Thus

$$\ell'(K) \leq \frac{\text{Area}(K'' \setminus K')}{\text{Area}(L)} \quad \text{and} \quad s'(K) \leq \frac{\text{Area}(K'' \setminus K')}{\text{Area}(S)}.$$

Now,

$$\text{Area}(K'' \setminus K') = \text{Area}(K'' \setminus K) + \text{Area}(K \setminus K').$$

The area of $K \setminus K'$ is bounded by φP. The area of $K'' \setminus K$—the white tube in (3.35)—is φP plus the sum of the areas of the circular sectors. Since the total turn angle around the boundary of a polygon is 2π (here by "turn angle" at a vertex we mean π minus the internal angle), the total area of the angular sectors is $2\pi\varphi$, and this proves (3.34). ∎

We see from Lemmas 3.35 and 3.36 that, in a Robinson tiling, the number of tiles (of each type) inside a compact convex set grows linearly with the area of the set, while the number of tiles crossing the boundary grows at most linearly with the perimeter.

3.6 Final Comments

The standard reference for Robinson triangles is the unpublished paper [Rob75]. I would like to thank the Stanford University Archives for sending me a copy of the preprint from Martin Gardner's personal collection [OAC].

Robinson's paper is only four pages long, but it states (without proof) many of the results reported in this chapter, including the less known ones (e.g. that "any complete line in a tiling must be an axis of symmetry", and that there are uncountably many such tilings).

Apparently the only reference with proofs is the book [GS87]. However, one should note that even in [GS87] some proofs are only sketched or simply missing (e.g. [GS87, 10.5.9] or [GS87, 10.5.11]), and sometimes some details are overlooked. For example, it is not true that if a topological disk contains only one vertex of a Robinson tiling, then it is entirely contained in the vertex neighborhood (cf. the proof of [GS87, 10.5.4], and compare it with our proof of Lemma 3.20 and Proposition 3.21).

It is also not true (see e.g. [Con94, p. 180, last three lines]) that, after enough Compositions, two points are always contained in the same tile. This claim is sometimes used to justify the statement in Proposition 3.27(ii). However, it is easy to find counterexamples: in the Sun or the Star tiling, two antipodal points will never be contained in the same tile, no matter how many Compositions we apply. What is true is that after enough Compositions, any two points (in fact, any bounded set) will be contained in a vertex neighborhood (as a consequence of Lemma 3.20).

A final comment on the protosets (3.1) and (3.5) is in order: while a reflection across the height transforms A-triangles into B-triangles and vice versa, an analogous reflection applied to (3.5) changes the protoset. The function transforming Robinson tilings into tilings with protoset (3.5) commutes with direct isometries, but not with reflections.

Chapter 4
Penrose Tilings

In this chapter we study the aperiodic protosets discovered by Penrose. Some of their properties will be deduced from analogous properties of Robinson triangles. Others will require completely new tools.

There are two famous sets of prototiles introduced by Penrose, each composed by two elements. The first set is given by a *kite* (K) and a *dart* (D):

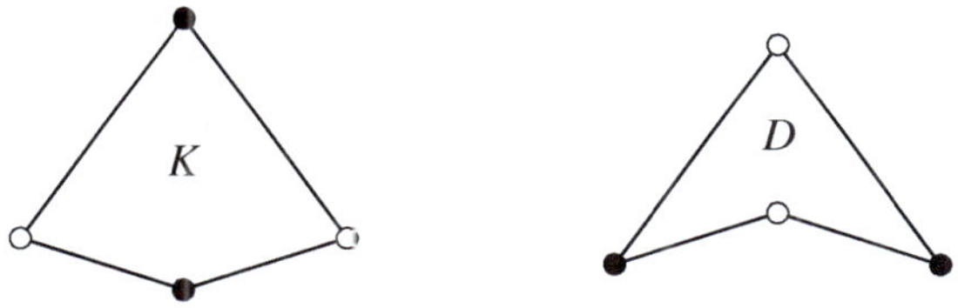

Here all the short edges have length 1, all the long edges have length φ, and the internal angle on the top vertex of both the kite and the dart is $2\pi/5$ (this fixes all the other angles).

The advantage of kites and darts is that one has only two prototiles, invariant under reflections, and with a very simple matching rule: two vertices can touch only if they are of the same color: both black or both white. The matching rule is designed to prevent the formation of a parallelogram:

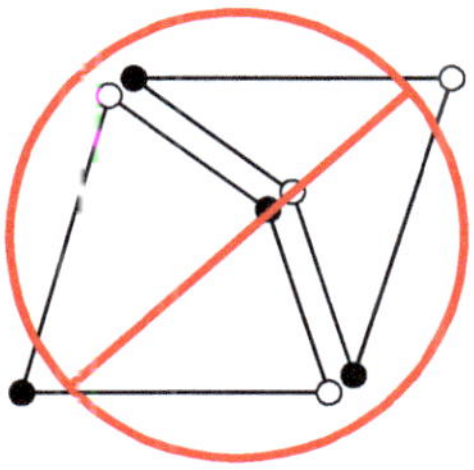

that could be easily used to tile the plane periodically.

F. D'Andrea, *A Guide to Penrose Tilings*,
https://doi.org/10.1007/978-3-031-28428-1_4

It is not difficult to enforce the same matching rule using arrows:

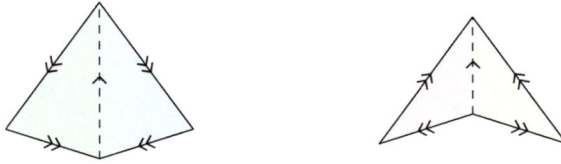

Even if the single arrow on the dashed line is not needed for the matching rule, if we draw it, it becomes evident the relation between kites and darts and Robinson triangles. Note that, in a Robinson tiling, every L_A (resp. S_A) tile has an L_B (resp. S_B) tile attached along the edge with a single arrow, and removing these edges we get a tiling by kites and darts.

Remark 4.1 From every partial tiling by kites and darts we can obtain a partial tiling by Robinson triangles "cutting" kites and darts in half. From every tiling of the whole plane by Robinson triangles, we get a tiling by kites and darts by removing all the edges with a single arrow.

A second set of prototiles is given by *Penrose rhombi*:

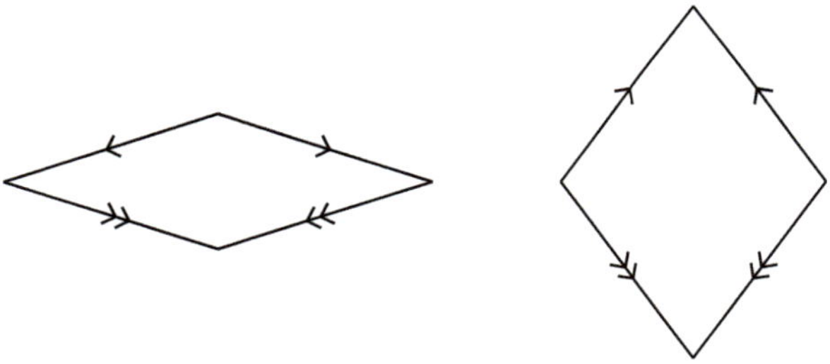

Here all the edges have length φ. In the rhombus on the left in the picture (the *thin* one) the angles are $\pi/5$ and $4\pi/5$, in the rhombus on the right (the *thick* one) the angles are $2\pi/5$ and $3\pi/5$.

The nice thing about Penrose rhombi is that there is only one length involved, and they are a really minimal modification of square tilings.

There is a substitution rule transforming each partial tiling by Penrose rhombi into one by Robinson triangles:

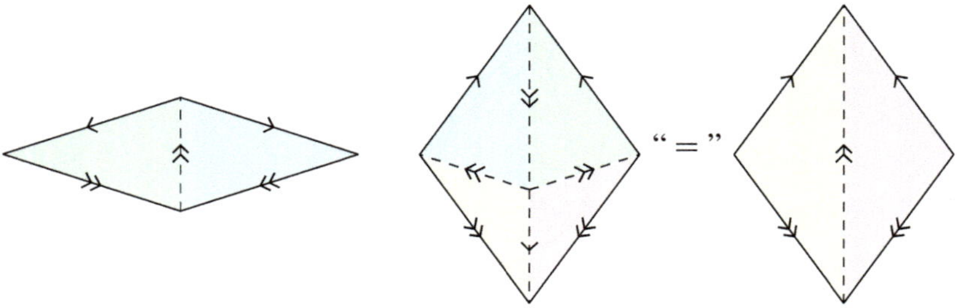

In the previous picture it is also illustrated how to pass from Penrose rhombi to the protoset $\{L_A, L_B, \varphi S_A, \varphi S_B\}$ in (3.14). For obvious reasons we refer to the triangles in this protoset as *half rhombi*.

Every tiling of the plane by Robinson triangles can be transformed into one by half rhombi using Composition 1. And in a tiling of the plane by half rhombi every L_A (resp. φS_A) tile has an L_B (resp. φS_B) tile attached along the base. Removing the bases of all isosceles triangles, we get a tiling by Penrose rhombi.

Remark 4.2 From every partial tiling by Penrose rhombi we can obtain one by Robinson triangles covering the same set. From every tiling of the plane by Robinson triangles we get a tiling by Penrose rhombi using first Composition 1 and then removing the bases of all isosceles triangles.

The markings on the rhombi are uniquely determined by the vertex where the two double arrows meet. We could then forget about oriented edges and use rhombi with a marked vertex. However, it is harder to express the matching rule in terms of dots.

One more way to obtain the same matching rule is by using decorations:

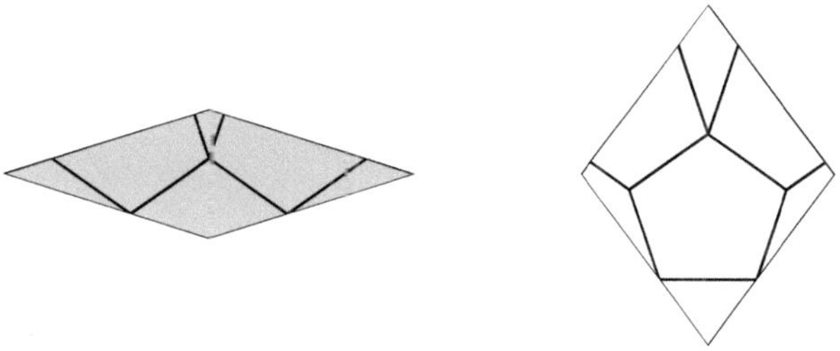

Here the rule is that edges must be joined in such a way that the decorations form straight-line segments.

Let us explain the origin of such decorations. In Fig. 4.1 we see how the first decorated rhombus is constructed: start with three pentagons like in the picture; draw the line ℓ passing through the points a and d; draw the line ℓ' passing through the points b and c; draw the line parallel to ℓ passing through e; draw the line parallel to ℓ' passing through f; the region bounded by the four lines is our first rhombus, and the portions of pentagons inside that region form the decorations. In Fig. 4.2 we see how the second rhombus is constructed: start again with three pentagons; draw the line ℓ passing through the points c and e; draw the line ℓ' passing through the points d and f; draw the line parallel to ℓ passing through b; draw the line parallel to ℓ' passing through a. The region bounded by the four lines is our second rhombus, and the portions of pentagons inside that region form the decorations.

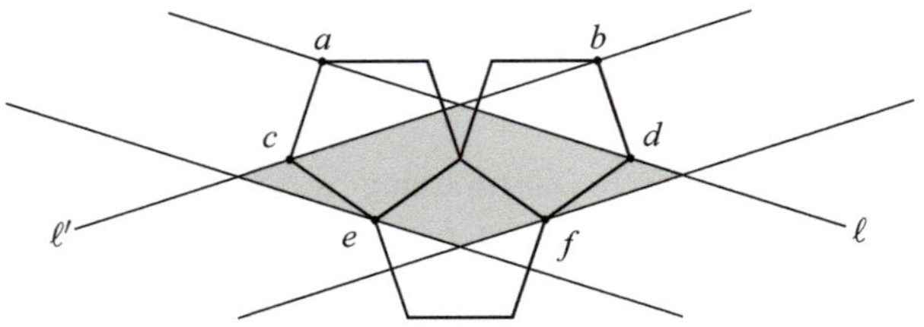

Fig. 4.1 Thin rhombus

Fig. 4.2 Thick rhombus

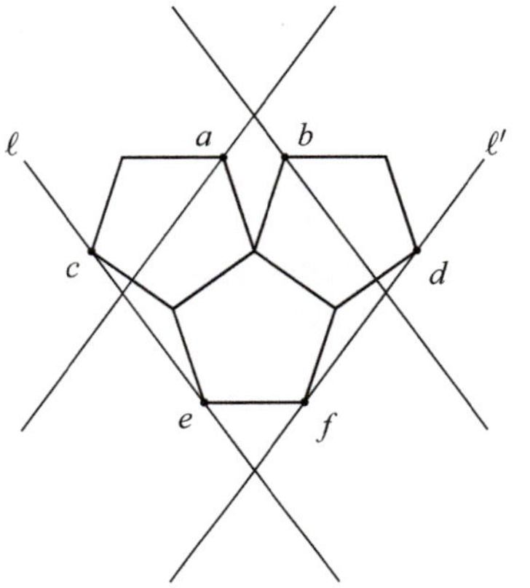

Now, if we start drawing a tiling with decorated rhombi, we will see several figures appearing in the backgroung:

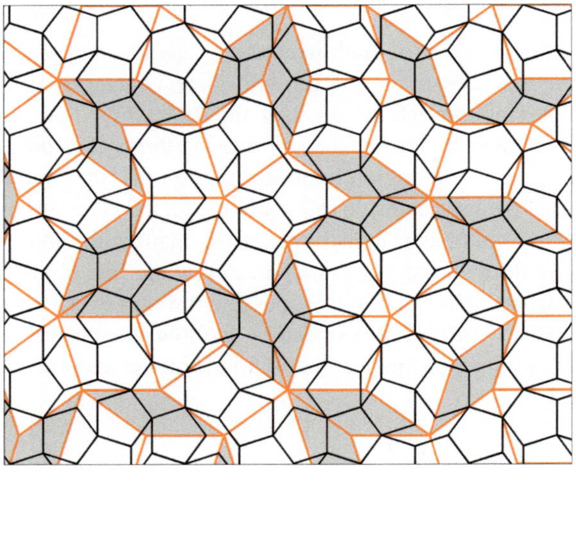

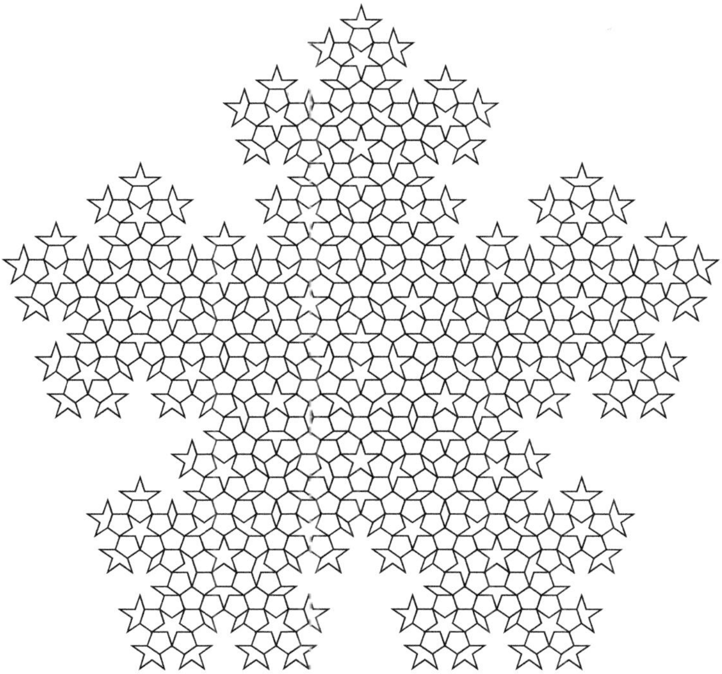

Fig. 4.3 Penrose snowflake

More precisely, we see a pentagon, a pentagram, a diamond and a boat:

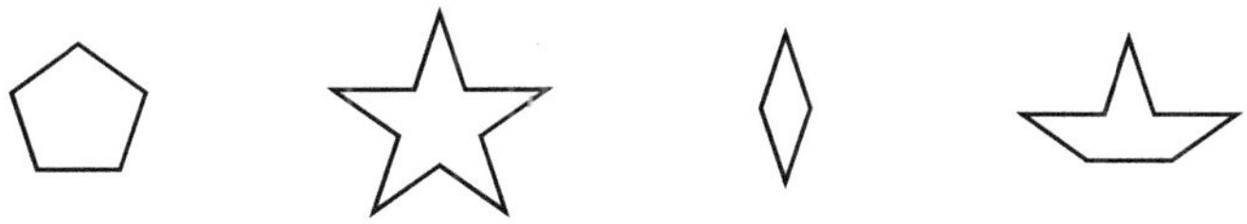

(4.1)

The decorations show how to pass from a tiling by Penrose rhombi to a tiling with protoset (4.1). One can similarly decorate (4.1) in such a way that from a tiling with this protoset it is possible to obtain a tiling by Penrose rhombi.

Historically, the set (4.1) with suitable matching rules (in fact, there are actually three pentagons in the protoset, with different matching rules) is the first example of aperiodic protoset discovered by Penrose, in an attempt to find a solution to Open Problem 2.13 (although here only two prototiles out of four have 5-fold symmetry). The interested reader can find more about this protoset in [GS87, Chap. 10].

With the prototiles (4.1), using the substitution rule illustrated e.g. in [GS87], one can produce partial tilings that look very much like self-similar fractals (but since we apply a scaling at each step, we get a tiling instead). Compare Fig. 4.3 with the first iterations of the Koch snowflake (see e.g. [Bur94]).

4.1 Aperiodicity, Classification, Local Properties, Symmetries

Consider as a protoset either the one given by a kite and a dart, or the one given by Penrose rhombi. Then:

(1) there exists tilings of the plane,
(2) both protosets are aperiodic,
(3) the set of equivalence classes of tilings has the cardinality of the continuum,
(4) there are exactly two tilings (up to congruence) with 5-fold symmetry,
(5) there are infinitely many inequivalent tilings with reflection symmetry,
(6) every patch of a tiling appears infinitely many times in every other tiling (with the same protoset).

These properties follow immediately from Remarks 4.1 and 4.2, from the properties of Robinson triangles in Chap. 3, and from the fact that the rules to transform Robinson tilings into Penrose tilings and back commute with isometries.

From Lemma 3.14 we deduce that:

Lemma 4.3 *In a tiling of the plane by Penrose rhombi, two thin rhombi can never share an edge with a single arrow.*

We can now list all possible vertex neighborhoods in a Penrose tiling. Each vertex neighborhood appears infinitely many times in every tiling of the plane (with the appropriate protoset).

Proposition 4.4 *In a Penrose tiling by kites and darts, only the following seven vertex neighborhoods are possible (cf. [GS87, Fig. 10.5.3]):*

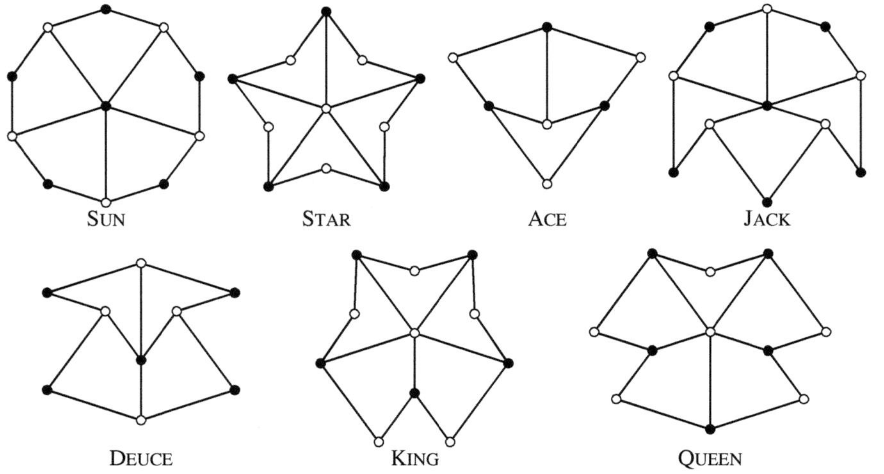

Proof The correspondence between Robinson tilings and Penrose tilings by kites and darts transforms vertices into vertices (no new vertices are created passing from one tiling to the other, in either direction). It is then sufficient to see what happens to the seven vertex neighborhoods in Proposition 3.9 when we remove the edges with a single arrow. One immediately sees that the wheel becomes the SUN; the star remains a STAR; the diamond is contained in an ACE (Lemma 3.12), that in the case of kites and darts is a vertex neighborhood. In the ice cream, the two S triangles have other S triangles attached along the edge with a single arrow, to form a dart; once we remove the edges with a single arrow we obtain the JACK. Similarly, the big kite becomes a DEUCE. In the cat, each L triangle has an additional L triangle attached along the edge with a single arrow, to form a kite, and it becomes the KING. Similarly, the shield becomes the QUEEN ■

In a tiling by Penrose rhombi there are eight types of vertex neighborhoods, reflecting the number of vertex neighborhoods in tilings with protoset (3.14). In the next proposition, we draw the arrows only on the edges incident with the central vertex: this fixes the decorations on all the other edges. Confront this with [dB81, Fig. 7] (one should note that here we adopt a different convention for the arrows: in [dB81] all single arrows are reversed). Vertex neighborhoods are labeled from V1 to V8 for future reference.

Proposition 4.5 *In a tiling of the plane by Penrose rhombi, only the following eight types of vertices are possible:*

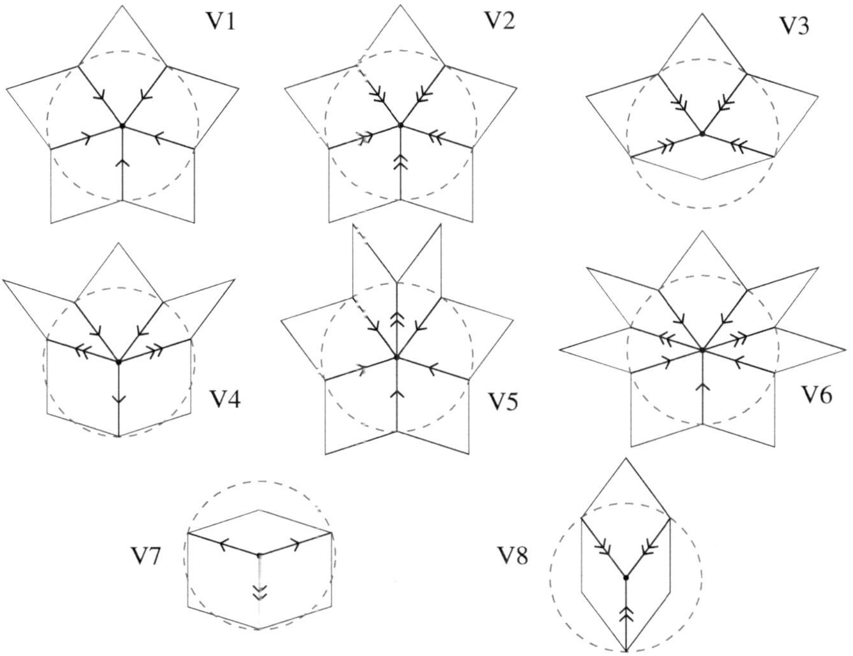

Proof We start again from Proposition 3.9. However, when passing from Robinson triangles to rhombi some vertices may disappear (we have to apply Composition 1 and then remove the bases of all the triangles). In other cases, there could be more than one way to complete a vertex neighborhood in Proposition 3.9 into one with Penrose rhombi.

The diamond becomes a single tile in a tiling with Penrose rhombi: the central vertex disappears. Every other neighborhood in Proposition 3.9, except the wheel, can be completed in a unique way to a vertex neighborhood with rhombi, as shown below.

In a star, every dart is contained in a diamond, which becomes the vertex neighborhood V2 in the above picture. Similarly, the shield becomes the vertex neighborhood V8 and the cat becomes the vertex neighborhood V3.

In a big kite, we see in the upper part two halves of thick rhombi: we get the vertex neighborhood V7 (rotated by 180°).

In the ice cream, we see in the lower part two halves of thick rhombi. Because of Lemma 4.3, the only way to complete the four top L triangles to rhombi is as illustrated in the vertex neighborhood V4.

Finally, by experimenting with all possibilities, using also Lemma 4.3, one shows that the wheel can become either one of the remaining three vertex neighborhoods in the picture. ■

Proposition 4.6 *A tiling of the plane by Penrose rhombi is uniquely determined by the position of its vertices.*

Proof Suppose we know the position of the vertices. We now show how to reconstruct (uniquely) the tiling from the vertex set.

For starters, observe that in a tiling by Penrose rhombi two vertices are connected by an edge if and only if they are at distance φ (the "if" part is the nontrivial one). To prove this, look in each vertex neighborhood in Proposition 4.5 at the region inside the red dashed circle. This circle has radius φ, and we see that the central vertex is connected by an edge to every other vertex on this circle, and to no other vertices. There are empty regions inside some circles, of the form:

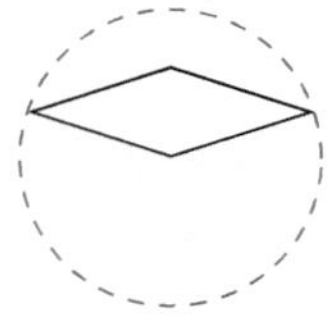

but there is no way to add tiles to the white region in such a way that more vertices appear on the dashed circle.

Starting from the vertex set, if then we join all pairs of vertices at distance φ, we reconstruct the tiling up to the markings on the edges, that can be reconstructed as follows.

We can distinguish almost all vertex neighborhoods even if we remove the arrows from the edges, simply by looking at the number of incident edges and at the angles between them, with the only exception given by V1 and V2. The first two vertex neighborhoods, see the next picture,

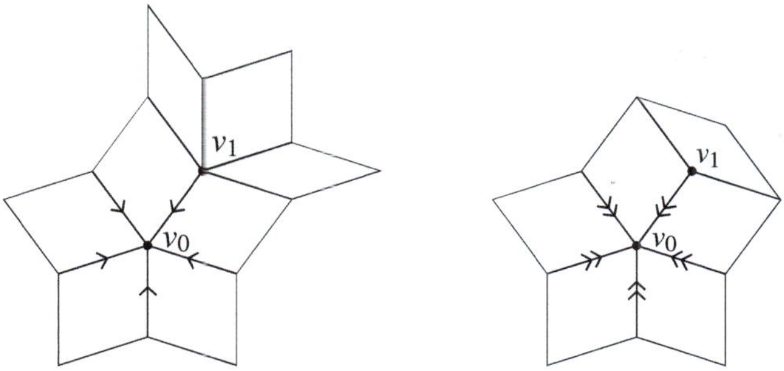

(4.2)

can be distinguished as follows. If the vertex neighborhood of v_0 is V1 (picture on the left), then the vertex neighborhood of v_1 must be V4. If the vertex neighborhood of v_0 is V2 (picture on the right), then the vertex neighborhood of v_1 must be V7. We can distinguish between the two neighborhoods, then, by looking at the tiles attached to the boundary edges.

If we know the vertices and the edges of our rhombus tiling, we can then look at each vertex, identify the vertex neighborhood, and then deduce the markings on the incident edges. ■

The proof of the previous proposition illustrates how to reconstruct a tiling of the plane by Penrose rhombi from its vertex set. An example is in the picture below:

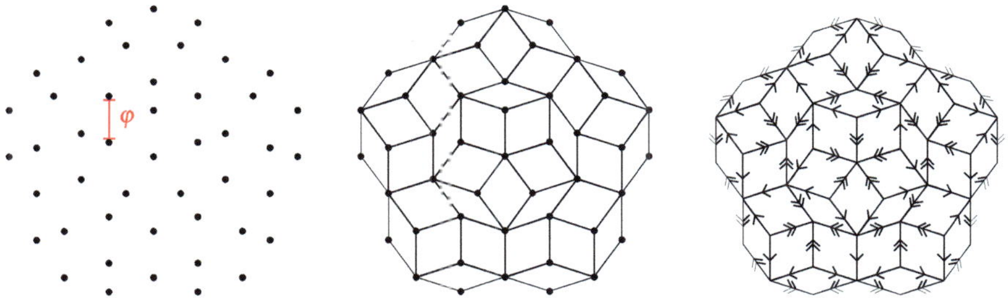

On the left we see the vertex set and, in red, a segment of length φ (to fix the scale of the picture). After joining all pairs of vertices at distance φ with an edge, we get the picture in the middle. Finally, a study of all vertex neighborhoods gives the picture on the right.

A last result that can be deduced from the previous chapter is that both tilings by kites and darts and Penrose rhombi are metrically prototile balanced. Let us start with kites and darts.

Proposition 4.7 *Let $(K_n)_{n\in\mathbb{N}}$ be a sequence of non-empty compact convex subsets of $\mathbb{R}^2$ such that $\lim_{n\to\infty} A_n/P_n = \infty$, where A_n is the area and P_n is the perimeter of K_n. Then, for any Robinson tiling of the plane:*

$$\lim_{n\to\infty} \frac{\#\text{ kites inside } K_n}{\#\text{ darts inside } K_n} = \varphi.$$

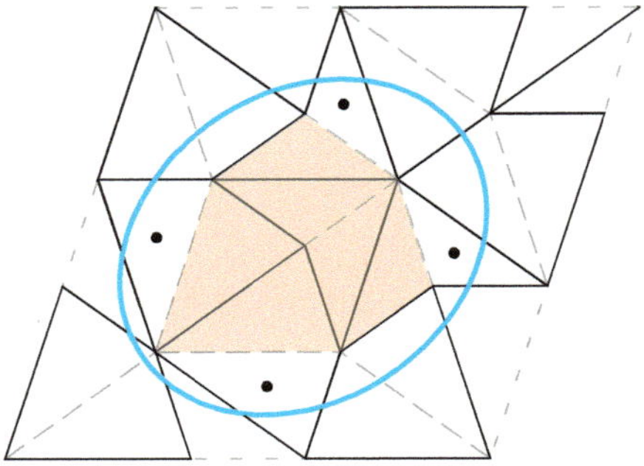

Fig. 4.4 Proof of Proposition 4.7

It is a common mistake to believe that this follows immediately from Proposition 3.34 and the decomposition of kites and darts into triangles. In particular, it is usually claimed that since every kite is formed by two L triangles and every dart by two S triangles, then for every compact convex set K one has (provided the denominator is not zero):

$$\frac{\text{\# kites inside } K}{\text{\# darts inside } K} = \frac{\not{2}}{\not{2}} \cdot \frac{\#L \text{ triangles inside } K}{\#S \text{ triangles inside } K}. \tag{4.3}$$

This is obviously not true, as shown in Fig. 4.4.

In the above example, if K is the ellipse whose boundary is drawn in blue, then the left hand side of (4.3) is 0 (there are no kites and one dart inside K), while the right hand side is $1/2$. The problem is that there are some triangles inside K that, when completed to form kites and darts, cross the boundary. This can be fixed as follows.

Proof of Proposition 4.7. Given $K \subset \mathbb{R}^2$ and a tiling of the plane by kites and darts, denote by $\ell(K)$ the number of L triangles inside K and by $s(K)$ the number of S triangles. We want to relate these two numbers to the number of kites and darts.

In Fig. 4.4 we see that in order to compute the number of kites we should count the number of L triangles inside K and subtract the number those crossing the boundary of K, that are marked with a bullet in the picture. The same for darts.

Denote by $\ell'(K)$ (resp. $s'(K)$) the number of L triangles and (resp. S triangles) that are crossed by the boundary of K.

Then, the correct version of (4.3) is (assuming that the denominators are not zero):

$$\frac{\ell(K) - \ell'(K)}{s(K)} \leq \frac{\text{\# kites inside } K}{\text{\# darts inside } K} \leq \frac{\ell(K)}{s(K) - s'(K)}.$$

From Proposition 4.7, $\ell(K)/s(K) \to \varphi$. From Lemmas 3.35 and 3.36:

$$\frac{\ell'(K)}{\ell(K)} \leq c' \frac{P + \pi}{A} \frac{A}{\ell(K)} \longrightarrow c' \cdot 0 \cdot c = 0$$

and

$$\frac{s'(K)}{s(K)} \leq c'\varphi \frac{P+\pi}{A}\frac{A}{s(K)} \longrightarrow c'\varphi \cdot 0 \cdot c\varphi = 0,$$

hence the thesis. ■

The next proposition for Penrose rhombi can be proved similarly. Instead of half kites and half darts one has to use half rhombi. The results in Sect. 3.5 can be proved for the protoset (3.14) as well, with the only difference that now the golden gnomon is the one with bigger area. Then, the result for rhombi can be proved from the one for triangles in the very same way we did for kites and darts.

Proposition 4.8 *Let $(K_n)_{n\in\mathbb{N}}$ be a sequence of non-empty compact convex subsets of $\mathbb{R}^2$ such that $\lim_{n\to\infty} A_n/P_n = \infty$, where A_n is the area and P_n is the perimeter of K_n. Then, for any Penrose tiling of the plane by rhombi:*

$$\lim_{n\to\infty} \frac{\#\text{ thick rhombi inside } K_n}{\#\text{ thin rhombi inside } K_n} = \varphi.$$

4.2 Imperfect Substitution Rules

There is a substitution rule, for both kites and darts and rhombi, that can be deduced from the one for Robinson triangles. It is sometimes referred to as "imperfect substitution", because we don't simply rescale the tiles and fraction them, but we have to add some pieces to complete the picture. The substitution rule for rhombi is:

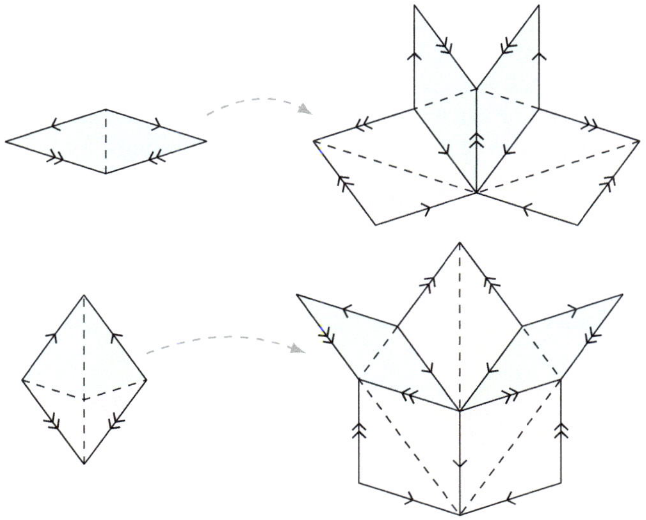

We draw the subdivision into triangles to better illustrate what is happening. One can see that the final patches are not simply inflated rhombi, but slightly bigger. The inflation and Decomposition of triangles transforms each tile in the protoset (3.1) into one or two triangles in the protoset (3.5), then one completes the picture by observing that in a tiling with protoset (3.5), each isosceles triangle has one of the same type attached along the base to form a rhombus.

When one attempts to apply this substitution rule to a partial tiling, or even to a single tile but multiple times, one will notice that the partial tilings obtained by substitution of adjacent tiles sometimes overlap. This is not a problem, of course, because the overlapping tiles will always match (and will therefore be considered the same tile).

Similarly one derives the substitution rule for kites and darts, that is given in the following pictures:

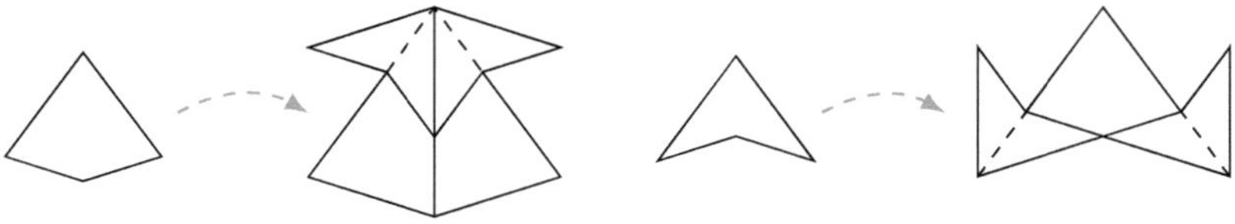

We see that the kite grows a hat, and the dart grows two arms.

In the other direction we have Composition, that takes a particularly simple form in the case of rhombi. Given a tiling by Penrose rhombi, we can cut the rhombi in half and transform it into one with protoset (3.14), the we apply Composition 2 followed by Composition 1 and we get a tiling with protoset similar to (3.14) but inflated by a factor φ (note the reversed order of Compositions), next we remove the bases of all isosceles triangles and get a tiling by Penrose rhombi inflated by a factor φ. We will refer to this operation as Composition of rhombi.

Recall that a tiling of the plane by Penrose rhombi is uniquely determined by its vertex set (Proposition 4.6). The next proposition characterizes Composition of rhombus tilings in terms of their vertex set.

With a slight abuse of language, let us call *type* of a vertex the type of its neighborhood, that is V1,…,V8 in the case of rhombi (Proposition 4.5).

Proposition 4.9 *Let $\mathcal{T}$ be a tiling by Penrose rhombi and $\mathcal{T}'$ the tiling obtained from $\mathcal{T}$ by Composition. Then, the vertex set of $\mathcal{T}'$ is obtained from the vertex set of $\mathcal{T}$ by removing all the vertices of type V4 and V7.*

Proof In the following picture we see in red the vertices that disappear after Composition:

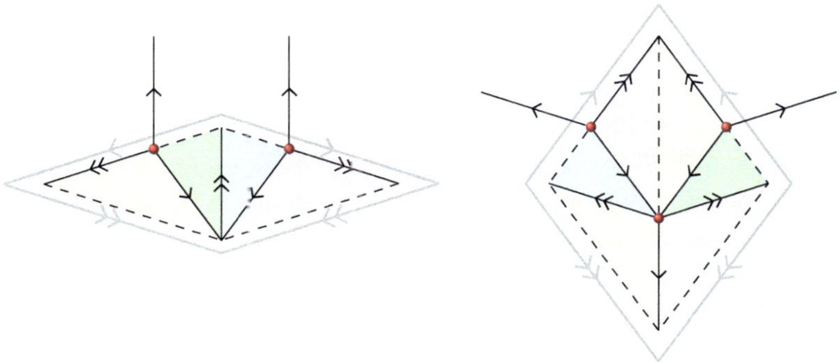

The central vertex in the picture on the right is of type V4. The remaining marked vertices are of type V7 (it is the only vertex neighborhood with more than one outgoing single arrow).

All other vertices remain (they are vertices in the big rhombi). None of the other vertices is of type V4 or V7: to see this, observe that they either have at least one ingoing double arrow, or at least three ingoing single arrows. ■

An example of Composition of Penrose rhombi is in Fig. 4.5.

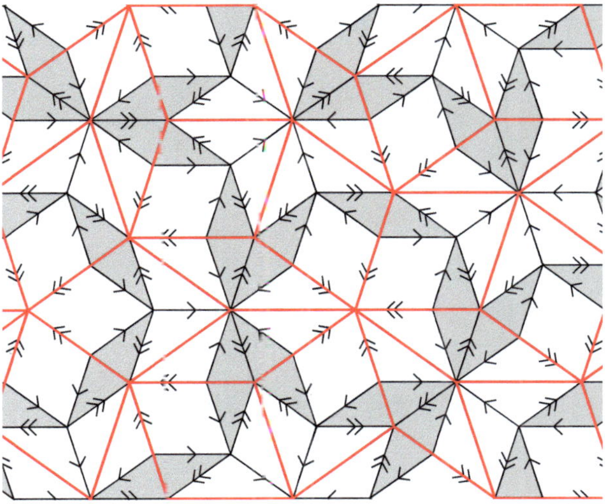

Fig. 4.5 Composition of Penrose rhombi

4.3 Conway Worms

Of course, all the tilings in Sect. 3.1 can be transformed into tilings by kites and darts. Of particular interest is the Cartwheel (Sect. 3.1.1). A patch is in Fig. 4.6.

At the center one may notice, in red, an ace: the starting point of the iterated inflation and substitution process. Outside the orange part we see a patch that, at a first look, seems to have a 10-fold symmetry. With a closer look, however, one realizes that this is not the case.

In green we see two patches known as short and long *bow tie*:

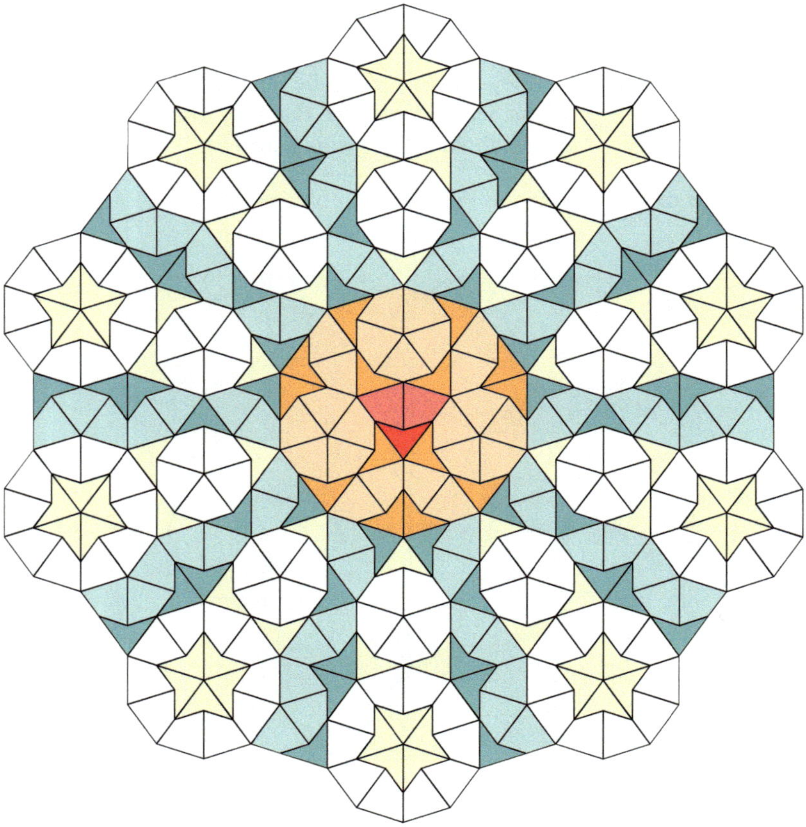

Fig. 4.6 The Cartwheel

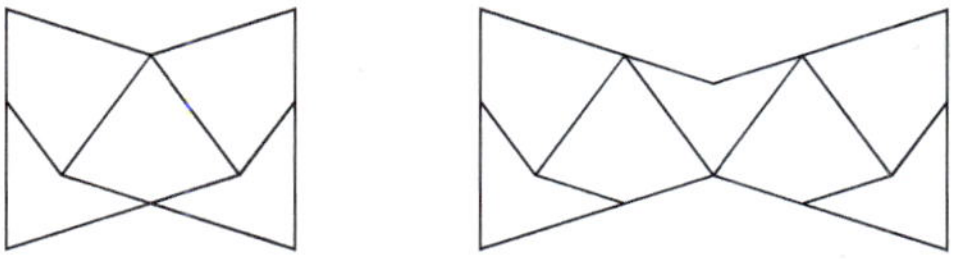

Short bow tie Long bow tie

The long bow tie can be obtained from the short one by cutting it in half and inserting a dart and two half-kites:

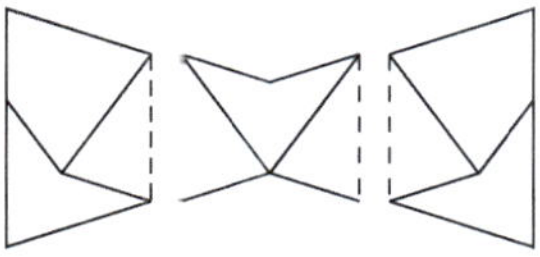

A sequence of bow ties placed end to end is what Conway called a *worm* [Pen89]. The next picture shows the result of the substitution rule applied to the two bow ties:

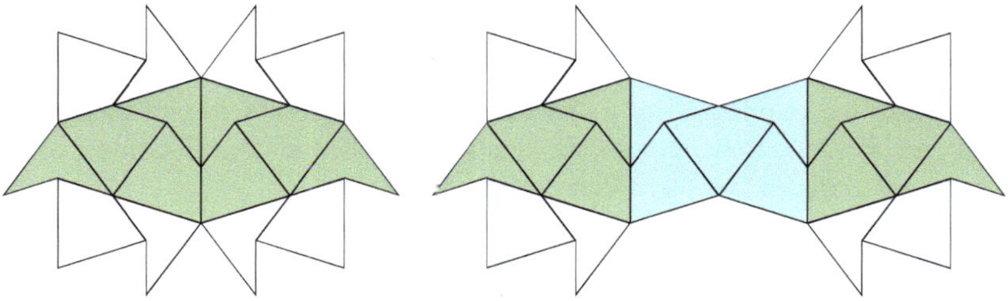

On the left, in green we see two halves of a long bow tie. On the right, two halves of a long bow tie in green, and a short bow tie in blue. It follows that, when we apply the substitution rule to a Conway worm, a longer worm is formed. In particular:

Proposition 4.10 *The Cartwheel contains ten Conway worms of infinite length radiating from a central patch.*

We see in Fig. 4.6 that six of these worms terminate on the central patch, while four of them join up in pairs across the Cartwheel so as to form two worms extending indefinitely in both directions and crossing each other (Fig. 4.7).

As a consequence of Proposition 3.21.

Proposition 4.11 *Every tiling of the plane by Penrose kites and darts contains infinitely many arbitrarily long finite Conway worms.*

To simplify the discussion, from now on by an *endless* Conway worm we will mean a two-sided infinite one.

One can show by trial and error that the intersection of two worms can only be an ace, like in Fig. 4.7. In particular, if we consider the two lines cutting the worms in half, these can only intersect at 72° ($2\pi/5$). Any additional endless worm in the Cartwheel would cross at least two one-sided infinite worms radiating from the center. Since it cannot intersect both at 72°, we conclude that:

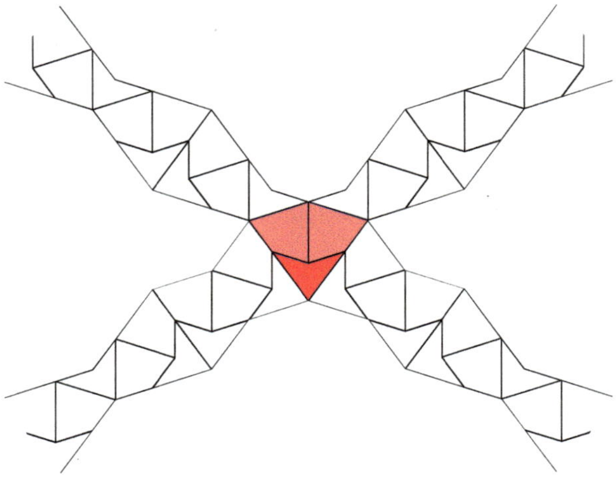

Fig. 4.7 Conway worms crossing

Proposition 4.12 *The Cartwheel contains only two endless Conway worms.*

If we take the two halves of a short bow tie and place them end to end, or the central part of a long bow tie, we can see how they change when passing from kites and darts to (triangles and then) Penrose rhombi:

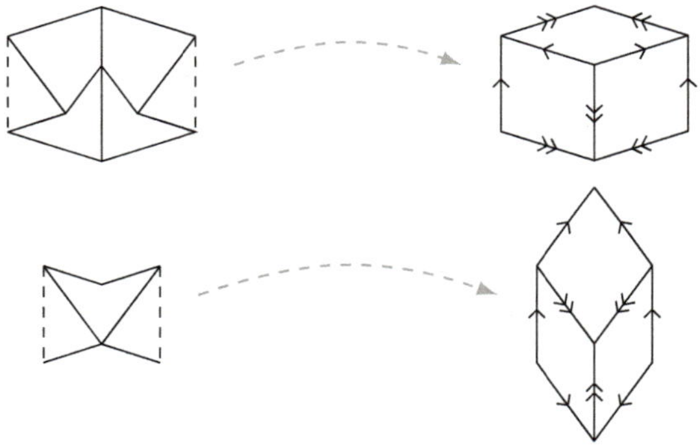 (4.4)

The second substitution is not obvious and follows from the fact that in a tiling of the plane every dart is contained in a diamond (and we get the thick rhombus on top), and that the two half-kites are contained in a bow tie. In particular, they are attached along the dashed line to thick rhombi. The only way they can form a rhombus is then as half of a thin one.

In a rhombus tiling, Conway worms become then sequences of non-regular hexagons (the last two vertex neighborhoods of Proposition 4.5):

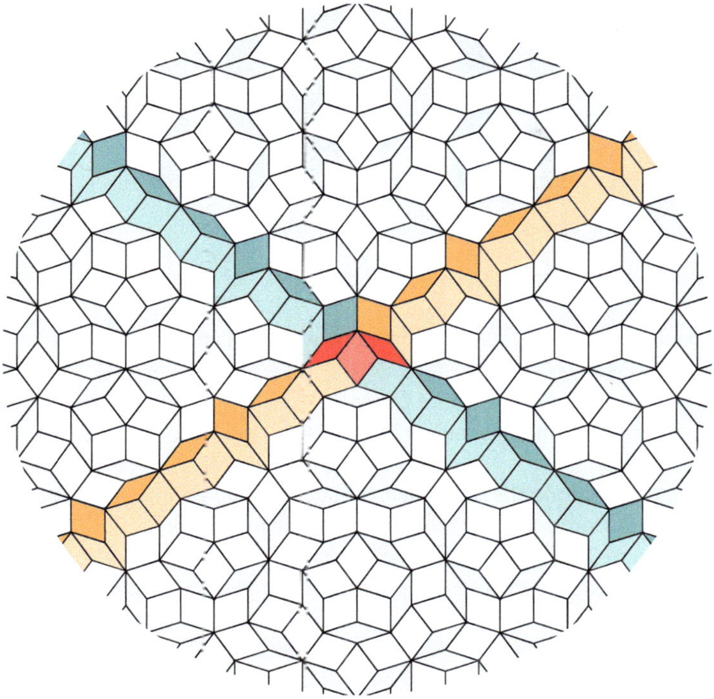

Fig. 4.8 Cartwheel with rhombi

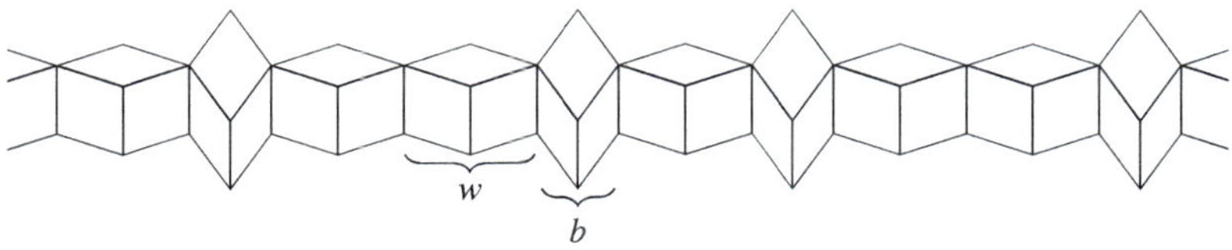

(4.5)

Figure 4.8 shows a portion of the Cartwheel made with rhombi, and the two endless worms crossing at the center.

4.3.1 *Musical Sequences*

A sequence of b's and w's is called a *musical sequence* if every b is followed by a w and there are no more than two consecutive w's. The reason for the name is that it reminds us of the black and white keys in a piano,

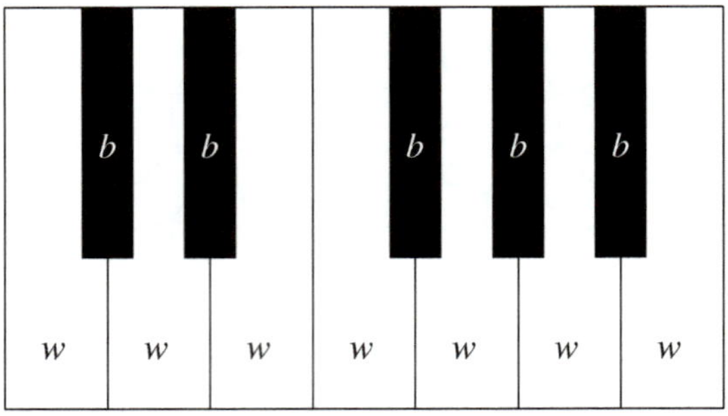

where every black key is always followed by a white one, and there are at most two consecutive white keys.

To any finite or one-sided infinite Conway worm made with rhombi we can associate a sequence of b's and w's: one w for each long hexagon, and one b for any short hexagon, cf. (4.5).

The same can be done for two-sided infinite worms, with a little care since there is no obvious starting point for the sequence. Given an endless worm, choose any two adjacent hexagons and number them 0 and 1. This fixes the starting point and orientation for counting hexagons in the worm. Then, we can consider the function $f : \mathbb{Z} \to \{b, w\}$ that, for each $n \in \mathbb{Z}$, associates to the nth hexagon his type, black or white. The function f is our doubly-infinite sequence.

Two such sequences f and g are *equivalent* if there exists n_0 such that either $f(n) = g(n + n_0)$ for all $n \in \mathbb{Z}$, or $f(n) = g(-n + n_0)$ for all $n \in \mathbb{Z}$ (i.e. they are related by a translation and possibly a reflection). For every endless Conway worm, the equivalence class of the associated sequence is uniquely defined (it does not depend on how we start numbering hexagons).

Proposition 4.13 *If a Conway worm with rhombi is a patch in a tiling of the plane, then the associated binary (b's and w's) sequence is a musical sequence.*

Proof We want to prove that in a legal Conway worm (finite or infinite) every short hexagon is followed by a long one, and there cannot be three consecutive long hexagons. For the first statement, look at the vertices v_1 and v_2 in the next picture:

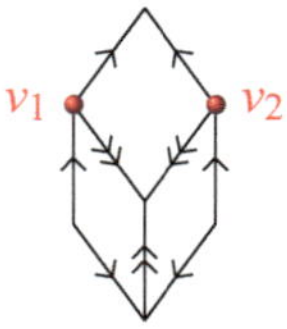

They both have one ingoing single arrow and one outgoing. For both vertices, the neighborhood must be V4 (Proposition 4.5), proving that the short hexagon has a long hexagon attached on each side.

For the second statement, consider the vertices v_1 and v_2 in the next picture:

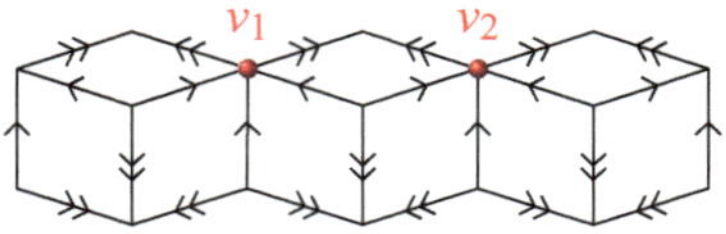

Both vertices have three ingoing single arrows and two outgoing double arrows. For both vertices, the vertex neighborhood can only be V6. But this is impossible because the two neighborhoods would overlap:

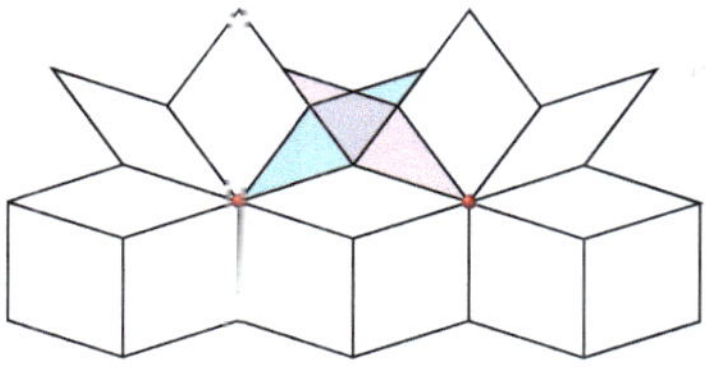

■

4.3.2 Hierarchical Structure

We saw that applying Decomposition to a worm leads to another worm. Now we show that endless worms are preserved under Composition as well. Half of the job is already done: if we apply Composition 1 to an endless worm made of bow ties, we get an endless worm made of non-regular hexagons, cf. (4.4). Now, let's apply Composition 2 to a Conway worm made of rhombi.

Suppose we have one or two w hexagons attached to two half b hexagons, like in the following pictures, on the left:

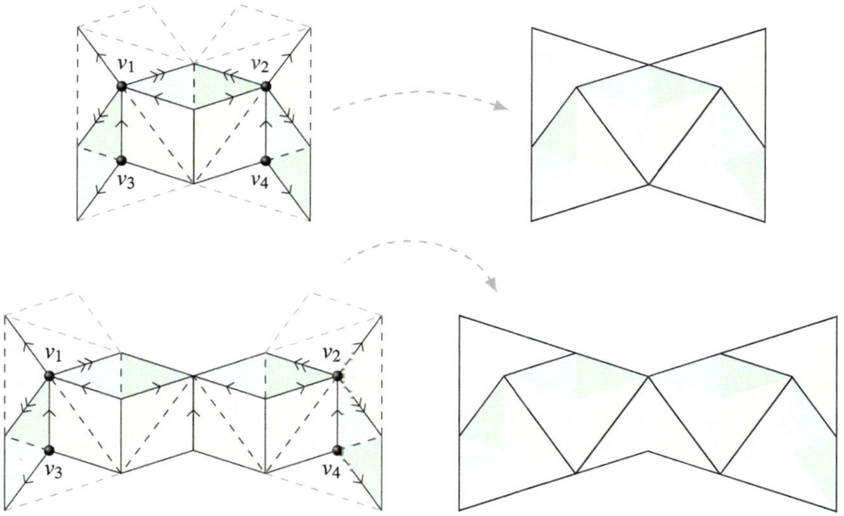

It follows from Proposition 4.13 that any endless Conway worm made of rhombi is a gluing of the two patches on the left of this picture. Observe that the vertex neighborhood of the vertices v_1 and v_2, in both pictures, is necessarily V4. This means that the white spaces on top are filled with thick rhombi, as illustrated by the dashed lines. On the other hand, v_3 and v_4 are necessarily contained in a big kite (Lemma 3.15). If we first remove from each L triangle the edge with a single arrow (Composition 2) and then in the resulting tiling we glue half-kites and half-darts, we get the short bow tie (first picture, on the right) and the long bow tie (second picture, on the right), respectively.

Observe that if we apply Composition to a Conway worm with kites and darts, the bow ties are shifted and turned upside down. On the other hand, if we apply Composition to a tiling containing two endless worms crossing, like in Fig. 4.7, the intersection of the two worms does not move (but it is turned upside down). This idea is at the basis of the proof of the next proposition.

Proposition 4.14 *If a tiling by kites and darts contains two endless Conway worms crossing each other, then it must be a Cartwheel.*

Proof Let $\mathcal{T}$ be a tiling by kites and darts containing two endless Conway worms crossing each other. We first observe that whenever two worms cross, they must have an ace in common. Without loss of generality, assume that the position of the ace is like in Fig. 3.2. For every $n \geq 1$, applying Composition $2n$ times to this tiling leads to new worms, crossing in a huge ace "centered" at the origin. Applying Decomposition $2n$ times gives us back the original tiling, which then contains a Cartwheel $\mathcal{C}_{2n+1}$ of order $2n+1$ (cf. Sect. 3.1.1). Thus $\mathcal{T} \supseteq \bigcup_{n\geq 0} \mathcal{C}_{2n+1}$. But since the latter union of patches tiles the whole plane, the set inclusion must be an equality and $\mathcal{T}$ must be a Cartwheel. ■

If we combine the above Composition 2, transforming sequences of hexagons in bow ties, with Composition 1 in (4.4), we see that Composition has the following effect on a worm made of rhombi:

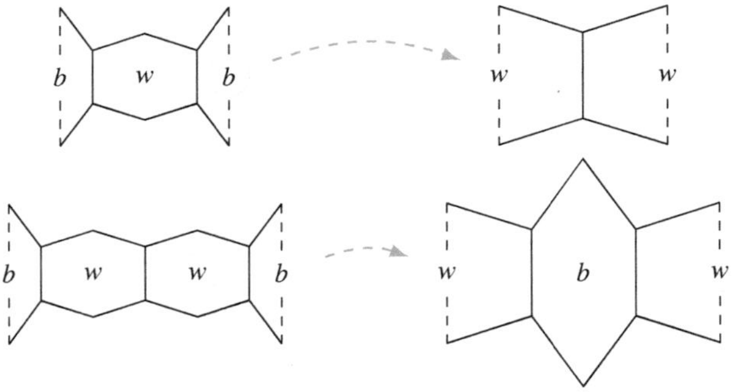

(4.6)

4.3.3 Fibonacci Strings

In terms of musical sequences, (4.6) corresponds to the following reduction rule for strings of b's and w's: every w that is not preceded by a b is replaced by a b, every block bw is replaced by w. We summarize this by the formulas:

$$bw \to w \qquad (\not{b})w \to b. \tag{4.7}$$

For example:

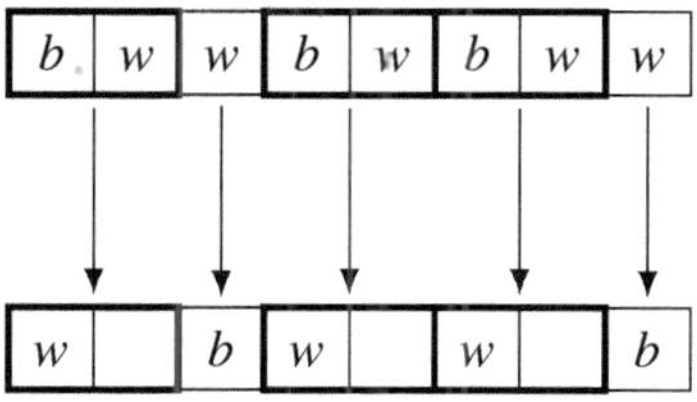

$= wbwwb$

The reduction rule (4.7) is well-defined on any (finite or infinite) string of b's and w's satisfying the condition that every b is followed by a w. In particular, it is well-defined on (finite or infinite) musical sequences.

Note that, when dealing with doubly-infinite sequences, it is important to keep track of the labeling of the letters in a string. Let us be more precise, then: in the reduction process, if there is a free slot in position $n \geq 0$, this will be filled with the first letter to the right; if there is a free slot in position $n < 0$, this will be filled with the first letter to the left. In the previous example, if the letters on top are labeled e.g. from -4 to 3, for the letters on bottom we get the following labeling:

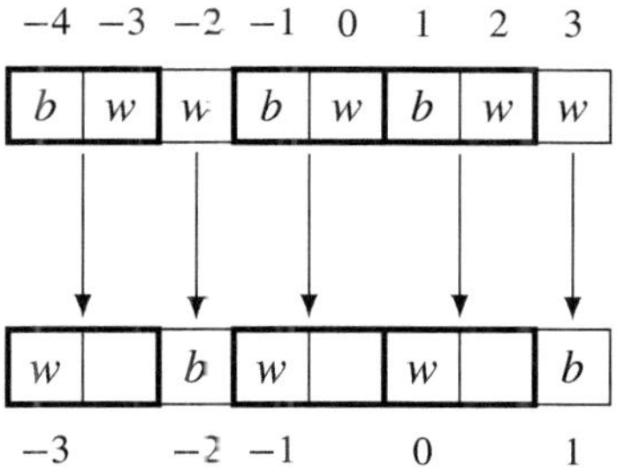

We can keep track of the numbering simply drawing a vertical bar between positions -1 and 0. The last example then becomes:

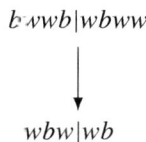

$$bwwb|wbww$$
$$\downarrow$$
$$wbw|wb$$

If we apply (4.7) to a musical sequence, the result is not necessarily a musical sequence. For example any sequence containing three consecutive bw's

$$\dots bwbwbw \dots,$$

becomes a sequence with three consecutive w's

$$\dots www \dots.$$

Definition 4.15 ([Sen95, Definition 4.1]) A doubly-infinite sequence of b's and w's is called a *Fibonacci string* if it is a musical sequence and it remains a musical sequence after application of an arbitrary number of the above reductions.[1]

Proposition 4.16 *If an endless Conway worm with rhombi is legal, then the associated binary (b's and w's) sequence is a Fibonacci string.*

Proof This is a simple consequence of the fact that the sequence of a worm is a musical sequence, and that Composition transforms worms into worms. ■

4.3.4 Fibonacci Tilings

In this section we provide a more geometric realization of Fibonacci strings and show that endless worms behave like one-dimensional tilings. Rather than giving the general definition of tiling (in arbitrary dimension), here by a tiling of a line we shall simply mean a cover by closed segments whose interiors don't overlap.

Given a Conway worm, draw a line crossing the hexagons orthogonally to the edges:

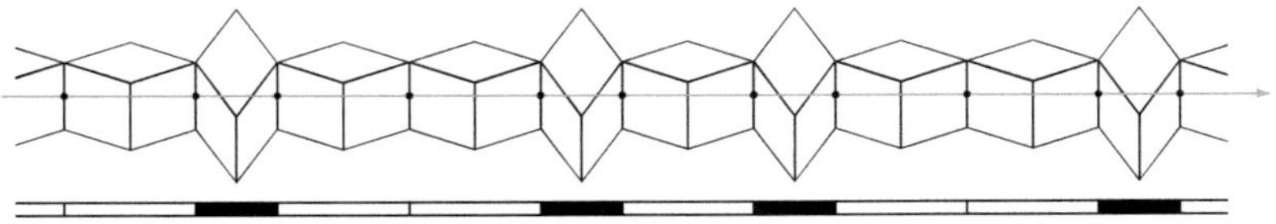

Fix arbitrarily an origin and an orientation on this line, so that we can identify it with $\mathbb{R}$. The intersection of a w hexagon with this line is a closed segment of length $2\varphi\cos(\pi/10) = \varphi\sqrt{\varphi+2}$. The intersection of a b hexagon with this line is a closed segment of length $2\varphi\cos(3\pi/10) = \sqrt{\varphi+2}$. The ratio of the two lengths is φ. In the figure above, long segments are depicted as thin white rectangles and short segments are thin black rectangles.

In this way, to every Conway worm we associate a one-dimensional tiling.

We can define composition and decomposition rules implementing those of Penrose tilings on worms in terms of segments. Composition replaces a black and white segment by a longer white segment, and changes color to every white segment that is not preceded by a black one:

Observe that if we start with any two segments whose length ratio is φ, we end with two segments whose length ratio is still φ.

[1] One should be aware that in some texts the names "musical sequence" and "Fibonacci string" are used interchangeably.

The advantage of one-dimensional tilings over strings is that the above composition rule commutes with one-dimensional translations and scalings (but not reflections). Note, however, that this composition rule does not reproduce exactly the composition of Penrose tilings restricted to worms: to reproduce the effect of Composition on worms one should move the corresponding one-dimensional tiling to the left of half big black segment after composition.

Given a protoset of two closed segments whose length ratio is φ, we can define musical and Fibonacci tilings in parallel to the definitions of the previous section. Denote by b the short and w the long segment, a *musical tiling* is a sequence of b and w segments, sharing only the endpoints, such that every b is followed by a w and there are no more than two consecutive w's. A *Fibonacci tiling* is a musical tiling of the line that remains a musical tiling after an arbitrary number of compositions.

Fibonacci strings can be generated from finite strings with the following substitution rule (Decomposition):

This is well-defined on arbitrary (partial) tilings by b's and w's, and transforms any tiling with no consecutive b's into a musical tiling. (This is easy to see: any tiling in the image of this map has b's always followed by w's, and a block www can only be generated by a bbb or by a wbb, hence the thesis.)

On Fibonacci tilings, where both are well-defined, Composition and Decomposition are inverse to each other.

If we start with a pair wb with common vertex at the origin and apply inflation by φ and then Decomposition, over and over again, we get:

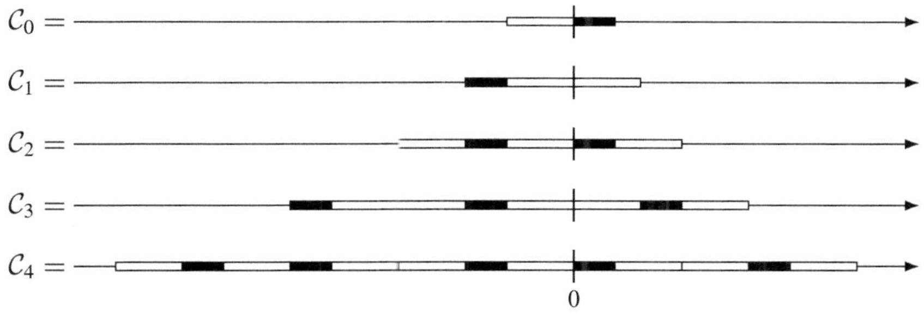

If start with a single tile, w or b, we generate a sequence of patches each one being the concatenation of the previous two, hence the name "Fibonacci". However, if we start with a wb, the sequence we generate doesn't have this property, suggesting that this name maybe is not completely appropriate.

Since $\mathcal{C}_0 \subset \mathcal{C}_2$, applying the substitution rule n times to this inclusion we prove that $\mathcal{C}_n \subset \mathcal{C}_{n+2}$, so that the two one-dimensional tilings

$$\mathcal{C}_{\text{even}} := \bigcup_{n\in\mathbb{N}} \mathcal{C}_{2n} \qquad \mathcal{C}_{\text{odd}} := \bigcup_{n\in\mathbb{N}} \mathcal{C}_{2n+1} \tag{4.8}$$

are well defined. Since $\mathcal{C}_0$ has no consecutive b's, each $\mathcal{C}_n$ with $n \geq 1$ is a musical tiling, which means that both $\mathcal{C}_{\text{even}}$ and $\mathcal{C}_{\text{odd}}$ are musical tilings (the condition for a tiling to be musical is "local").

Since composition transforms $\mathcal{C}_{\text{even}}$ in an inflated version of $\mathcal{C}_{\text{odd}}$, and vice versa, they both are Fibonacci tilings.

Several properties of Fibonacci tilings can be proved with arguments similar to those used in Chap. 3 for Robinson tilings.

For starters, every Fibonacci tiling is non-periodic. By contradiction, assume that a Fibonacci tiling is invariant under a translation of a parameter $t > 0$. After several compositions we arrive at a tiling still invariant under translation by t, but with tiles both longer than t, which is impossible.

A vertex neighborhood in this context is given by two adjacent segments, and there are only three possibilities: bw, wb, ww.

The tilings in (4.8) have the role of the Cartwheel. We can see that $\mathcal{C}_3$ contains all three possible vertex neighborhoods. In a Fibonacci tiling, every tile is contained in a block wb (every b is surrounded by w's, and each w is attached to at least one b). The block wb has the role of the ace. With the usual Composition/Decomposition trick, one proves that in a Fibonacci tiling: (1) for any $n \geq 1$, every tile is contained in a copy of $\mathcal{C}_n$; (2) every vertex neighborhood appears infinitely many times; (3) every patch of a Fibonacci tiling appears infinitely many times in every other Fibonacci tiling. Thus, Fibonacci tilings are all locally equivalent, and they are repetitive.

One can associate to every Fibonacci tiling an index sequence and show that there are uncountably infinitely many inequivalent tilings.

In particular, it follows that endless Conway worms are non-periodic, repetitive, and all locally equivalent.

4.4 Ribbons

In a tiling by Penrose rhombi, every Conway worm contains a "ribbon".

In general, suppose we have an edge-to-edge tiling of the plane with prototiles given by parallelograms. We will call it a *parallelogram tiling*, for short. Let T_0 be a tile and e_0 and e_1 two parallel edges. Call T_1 the tile sharing the edge e_1 with T_0 and e_2 the edge of T_1 parallel to e_1. Continue by induction: for every $n \geq 1$, call T_n the tile sharing the edge e_n with T_{n-1} and let e_{n+1} the edge of T_n parallel to e_n. Do the same with e_0 to get a sequence of tiles T_{-1}, T_{-2}, etc. in the opposite direction. In this way we get a doubly infinite sequence:

$$\ldots, T_{-n-1}, T_{-n}, \ldots, T_{-1}, T_0, T_1, \ldots, T_n, T_{n+1}, \ldots$$

of tiles sharing a bunch of parallel edges. Such a sequence is called a *ribbon*. (In fact, we don't care about the labeling of tiles: two sequences formed by the same set of tiles will be considered the same ribbon.)

Fig. 4.9 Ribbons

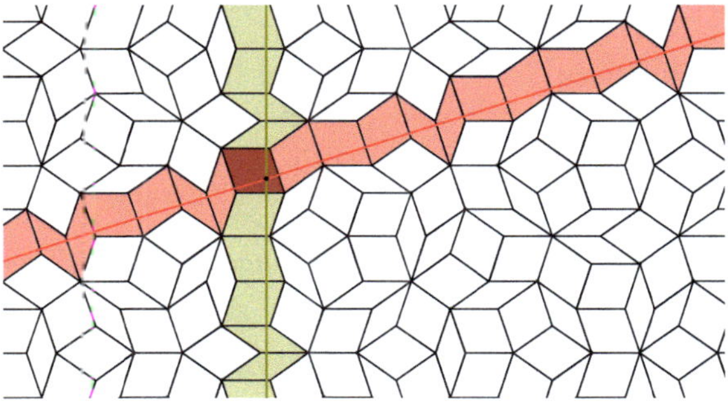

Fig. 4.10 Intersecting ribbons

An example in the case of Penrose tilings is in Fig. 4.9. Colored in the picture we see a number of vertical ribbons.

If the sequences of tiles defining two ribbons have non-empty intersection, we say that the two ribbons *intersect* or *cross*. In Fig. 4.9 we see a number of ribbons that do not cross; in Fig. 4.10 we see two intersecting ribbons (in different colors): in dark red we see the tile which is the intersection of the two ribbons. Observe that any tile is the intersection of exactly two ribbons.

Every ribbon has a well defined direction: in the notations above, to each ribbon one can associate the direction perpendicular to the edge e_0 (or to any edge shared by two consecutive tiles of the ribbon). In Fig. 4.10, over each ribbon we see a line (of the same color) in the direction of the ribbon. Two different ribbons with the same direction will be called *parallel*.

In a Penrose tiling, ribbons behave in some sense as lines. More precisely:

Proposition 4.17 *In a tiling of the plane by Penrose rhombi, two ribbons are parallel if and only if they have empty intersection.*

We will split the proof in two Lemmas, corresponding to the "if" and "only if" part. One implication is easy to prove and true for any parallelogram tiling. The other implication is a special property of Penrose tilings, as we shall see below.

Lemma 4.18 *In a parallelogram tiling, two parallel ribbons never cross.*

Proof Suppose that two distinct ribbons intersect. Their directions are then determined by the directions of the two pairs of parallel edges of the (any) common tile. But the four edges of a parallelogram (with positive area) cannot all be parallel. ■

In a general parallelogram tiling, two ribbons that don't cross need not be parallel. In the following picture we see two ribbons (in different colors) with no tile in common, which however have different directions:

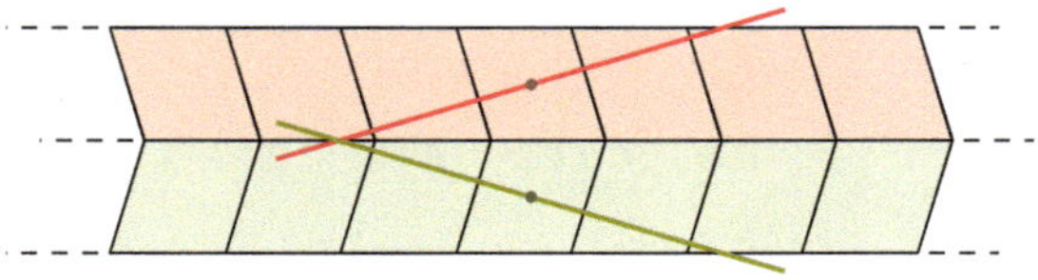

In the above picture we see that the ribbons form a strip whose boundary lines don't have, however, the same direction as the ribbon (given by the colored lines). The properties in the next lemma are special properties of Penrose rhombus tilings, and do not generalize to arbitrary parallelogram tilings.

Lemma 4.19 *In a tiling of the plane by Penrose rhombi:*

(i) every ribbon contains infinitely many rhombi of both types, thin and thick;
(ii) every ribbon is contained in a strip of width $\frac{3}{2}\varphi + 1 \approx 3.5$*, whose boundary lines have the same direction of the ribbon;*
(iii) there exists a line with the same direction of the ribbon that intersects all the tiles in the ribbon.

Proof Recall that the edges of the rhombi have length φ. Suppose we have a ribbon, and observe that in the edges shared by two consecutive tiles the single and double arrow are alternating (in both the thin and thick rhombus, opposite to a single arrow we always have a double arrow, and vice versa). We can assume that the direction of the ribbon is horizontal. We cut the ribbon along the edges with a double arrow. In this way we decompose the ribbon into a sequence of pieces, each formed by two tiles attached along an edge with a single arrow: we will refer to such a piece as a *domino* in this proof. Because of Lemma 4.3, up to a translation, there are only six types of domino pieces:

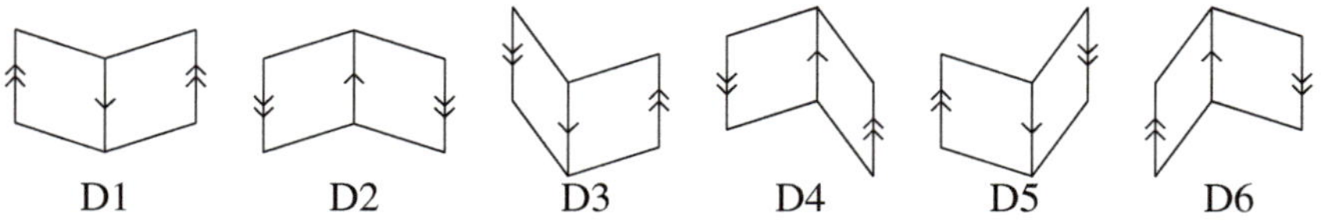

(Three of them are obtained from the other three by a 180° rotation.) These six pieces are copied, translated, and then attached along the vertical edges with double arrow to form the ribbon (respecting the matching rules).

Suppose, by contradiction, that the ribbon contains four pieces D2 attached together:

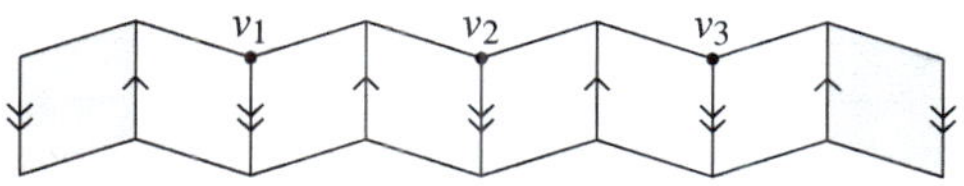

The vertex neighborhoods of the vertices v_1, v_2 and v_3 in the picture must be V7, and this is impossible (cf. Proposition 4.13). The same holds for D1. In the ribbon, after at most three consecutive pieces D1 or D2, there should be one of the pieces D3-D6 attached. In particular: there is at least one thin rhombus for every seven thick rhombi, and there is at least a thick rhombus for every thin rhombus (they always come in pairs in the pieces D3-D6). This proves (i).

Now, locate one of the pieces D3-D6 in the ribbon (we just proved that there must be at least one). Suppose it is D3 (the other cases being similar). Call T_0 the thick rhombus in this domino piece, and start counting tiles in the ribbon from there (with positive numbers on the right, and negative on the left).

Observe that D3 must be contained in a vertex neighborhood V7 (a non-regular hexagon); draw the six horizontal lines through the vertices of this hexagon, like in the following picture:

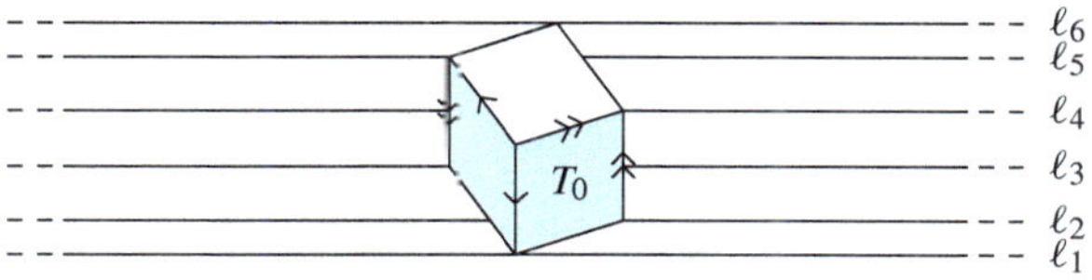

In the domino piece in the picture:

(∗) the vertical edge with double arrow pointing up has extremes on the lines ℓ_2 and ℓ_4; the vertical edge with double arrow pointing down has extremes on the lines ℓ_3 and ℓ_5.

Now we prove by induction that this is true for every domino piece in the ribbon; this in particular proves that the ribbon is entirely contained in the strip delimited by the lines ℓ_1 and ℓ_6.

We will prove the statement only for the half-ribbon to the right of T_0, the rest of the proof being analogous. Suppose by induction that, for some even $n \geq 0$, the domino piece formed by the pair tiles (T_{n-1}, T_n) satisfies the condition (∗) above. Attach one more domino piece to the right. If the right edge of T_n points up, we have the following three possibilities:

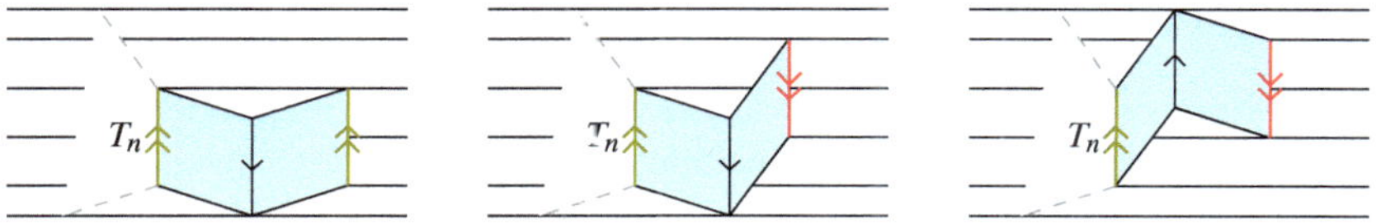

If it points down, we have the following three possibilities:

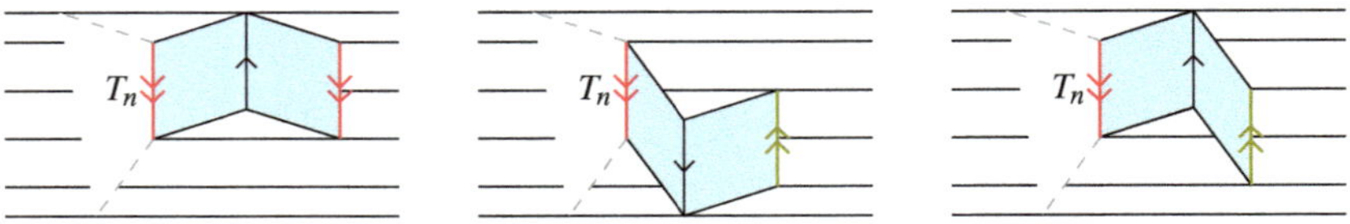

In every case, we see that the pair of tiles (T_{n+1}, T_{n+2}) satisfies the condition (∗).

To conclude the proof of (ii), we observe that the distance between ℓ_1 and ℓ_6 is

$$d(\ell_1, \ell_6) = \varphi\Big(1 + \cos\frac{\pi}{5} + \cos\frac{2\pi}{5}\Big).$$

From (A.6) we deduce that

$$d(\ell_1, \ell_6) = \varphi\Big(1 + \frac{\varphi}{2} + \frac{\varphi^{-1}}{2}\Big),$$

which combined with (A.4) gives $d(\ell_1, \ell_6) = \frac{3}{2}\varphi + 1$.

Finally, to prove (iii) we observe that any horizontal line between ℓ_3 and ℓ_4 intersects both tiles of the domino pieces, in all six cases. ■

Lemma 4.20 *In a tiling of the plane by Penrose rhombi, two ribbons with different direction have at least one tile in common.*

Proof Suppose we have two ribbons with different direction. Draw the strips containing the two ribbons, as in Lemma 4.19, and the polygonal chains connecting the midpoints of consecutive parallel edges of each ribbon. In the next picture the polygonal chains are in green and orange:

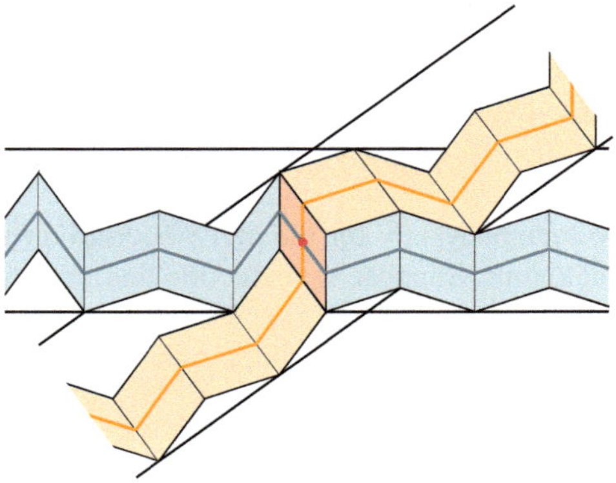

(4.9)

The two strips must intersect (their boundary lines are not parallel). Each polygonal chain connects opposite sides of the parallelogram which is the intersection of the two strips. We deduce that the two polygonal chains must intersect.

Each polygonal chain intersects only edges that are orthogonal to the direction of the ribbon surrounding it. Since the ribbons have different directions, there is no edge in the tiling that intersects both polygonal chains. In particular, the intersection point of the two chains must be in the interior of a tile. ■

Now, combining Lemmas 4.18 and 4.20 we get Proposition 4.17.

Using similar arguments, one could prove that two ribbons never cross more than once. This is, in fact, a general property of parallelogram tilings [FH13].

An interesting corollary of Proposition 4.17 is the following.

Corollary 4.21 *In a tiling of the plane by Penrose rhombi, ribbons can have only five different directions. These are given, up to a congruence, by the unit vectors* $e_n := e^{2\pi i n/5}$, $0 \le n \le 4$.

Proof Choose any ribbon R. Up to a rotation, we can assume that it has direction e_0. Let R' be any ribbon not parallel to R. It follows from Lemma 4.20 that R and R' have a tile T in common. Two edges of T are vertical, since the direction of R is horizontal. The other two edges give the direction of R':

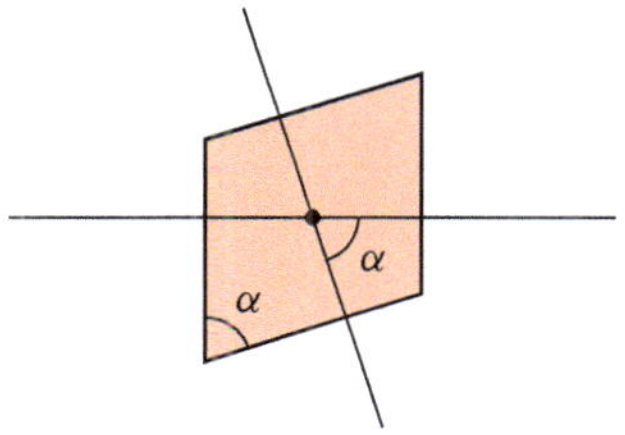

The acute angle α between the lines giving the direction of the ribbons is equal to the acute interior angle of the tile. We have only four possibilities: $\alpha = \pm\pi/5$, corresponding to the unit vectors e_3 and e_2, and $\alpha = \pm 2\pi/5$, corresponding to the unit vectors e_1 and e_4, respectively. ■

An immediate corollary is that:

Corollary 4.22 *In a Penrose tiling (by rhombi, triangles or kites and darts), each prototile occurs in a finite number of orientations.*

By this we mean that there is a finite subgroup Γ of $SO(2)$ such that, for each prototile T_0, every tile congruent to T_0 has the form $uT_0 + b$ with $u \in \Gamma$. In the example of Penrose tilings, Γ is the cyclic group of order 10 (each of the 5 possible directions corresponds to two orientations of the tiles).

The last corollary can be generalized to arbitrary parallelogram tilings (under the assumption that the protoset is finite), and more generally to protosets given by finitely many centrally symmetric convex polygons (since any centrally symmetric convex polygon can be tiled by parallelograms). For a proof, see [FH13].

The kite and domino tiling of Sect. 2.4 is an example of tiling where infinitely many orientations occur [BFG07].

The existence of finitely may orientations allows us to prove that Penrose tilings are repetitive. Given two congruent patches $\mathcal{P}$ and $\mathcal{P}'$, we will say that they have the same orientation if one is a translated copy of the other. Otherwise, we say that they have different orientations.

Proposition 4.23 *Every Penrose tiling is quasi-periodic.*

Proof Let $\mathcal{P} \subset \mathcal{T}$ be a patch in a Penrose tiling of the plane (by rhombi, kites and darts, or Robinson triangles). $\mathcal{T}$ contains infinitely many copies of $\mathcal{P}$. If they all have the same orientation of $\mathcal{P}$, the proof is complete. If not, let $\mathcal{P}'$ be a copy with different orientation. Now, the tiling contains infinitely many copies of $\mathcal{P} \cup \mathcal{P}'$. If they all have the same orientation, then we found also infinitely many copies of $\mathcal{P}$ with the same orientation, and the proof is complete. If not, let $\mathcal{P}'' \cup \mathcal{P}'''$ be a copy with different orientation. At least one of the components, say $\mathcal{P}''$, has orientation different from both $\mathcal{P}$ and $\mathcal{P}'$, and the patch $\mathcal{P} \cup \mathcal{P}' \cup \mathcal{P}''$ contains three copies of $\mathcal{P}$ with different orientation.

Continue with the same argument to find ten copies $(\mathcal{P}_i)_{i=0,\dots,9}$ of $\mathcal{P}$ with ten different orientations. The tiling contains infinitely many copies of the union $\bigcup_{i=0}^{9} \mathcal{P}_i$, and since there are only ten possible orientations, no matter how this union is rotated, at least one component will always have the same orientation as $\mathcal{P}$. ■

4.5 Non-locality and Empires

Recall that a partial tiling is called *legal* if it can be extended to a tiling of the plane.

Given a legal partial tiling $\mathcal{P}$, we say that a tile $T_0 \notin \mathcal{P}$ is *forced* by $\mathcal{P}$ if, for every tiling of the plane $\mathcal{T}$ extending $\mathcal{P}$, one has $T_0 \in \mathcal{T}$. The *empire* of $\mathcal{P}$ is the union of $\mathcal{P}$ and the set of all tiles that are forced by $\mathcal{P}$.

Lemma 4.3 gives us two simple examples:

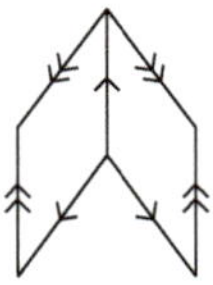

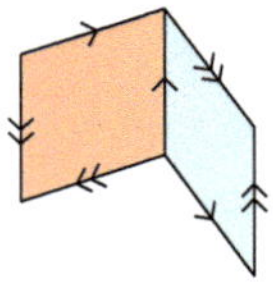

On the left we see an illegal partial tiling. On the right, the blue tile (thin rhombus) forces the red one (thick rhombus). Another example of forced tiles is in (4.2): there, the white vertex neighborhoods force the gray tiles.

In general, computing the empire of a legal tiling is a quite difficult task, since it is usually an infinite set. One can compute arbitrary large portions of an empire using *Ammann bars*. We will not discuss this topic here. One can see [GS87], where one can also find pictures of the empires of some kites and darts configurations. About ten years after the book was published, in the thesis [Min98] the author extended Conway and Ammann's work on forced tiles and noticed one case (the deuce's empire) where the above-mentioned method does not locate all forced tiles.

Observe that no finite partial tiling can force an entire tiling of the plane, because of the local equivalence of Penrose tilings: any finite legal partial tiling appears in every tiling of the plane with the same protoset. On the other hand, some infinite partial tilings can force an entire tiling of the plane, like in the case of endless Conway worms.

Proposition 4.24 *The empire of a legal endless Conway worm covers the plane.*

Proof Let $\mathcal{T}$ be a tiling of the plane by Penrose rhombi, and suppose it contains an endless worm. Choose a point p inside the worm. Composition transforms endless worms into endless worms (Sect. 4.3.2). Thus, after Composition, p is still inside a worm. Pass to Robinson triangles and assume that p is in the interior of a tile. If we use p as a base point to compute the index sequence of the tiling, we see that it depends only on the index sequence of the Fibonacci tiling associated to the worm. The worm uniquely determines, up to tail equivalence, the index sequence of $\mathcal{T}$, and then $\mathcal{T}$ itself. ■

One might believe that empires are connected (by this we mean that the union of tiles is a connected subset of $\mathbb{R}^2$). Below we provide a simple counterexample that can be worked out by hand.

Proposition 4.25 *Let $\mathcal{P}$ be the vertex neighborhood V4 of Proposition 4.5 (in blue in the next picture). The empire of $\mathcal{P}$ is disconnected. In particular, the red tiles in (4.10) are forced by $\mathcal{P}$:*

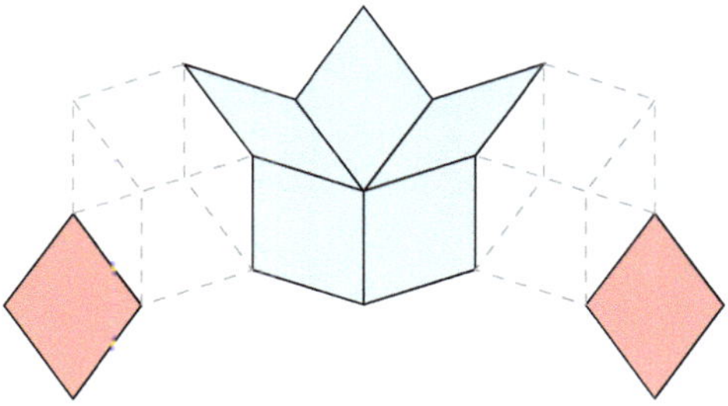

while the dashed regions don't belong to the empire.

Proof Let us prove that the red tile on the left is forced by $\mathcal{P}$.
Consider the following vertex v_1:

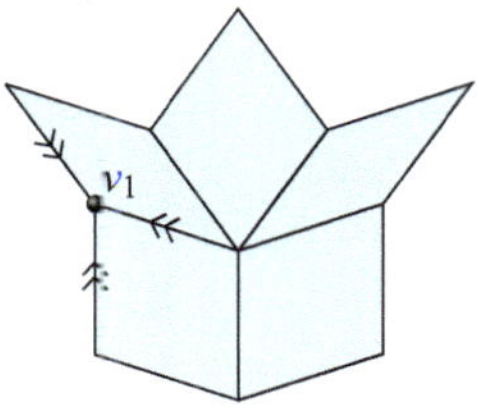

If we look at the list of vertex neighborhoods in Proposition 4.5, we see that the neighborhood of v_1 can only be V3 or V8. In the first case, there are two thick rhombi attached to v_1 like in the picture on the left, in the second case, a thin rhombus like in the picture on the right:

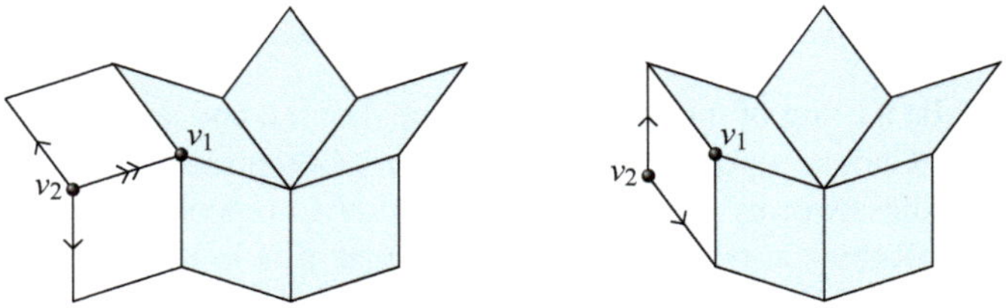

We now look at all possible vertex neighborhoods of v_2 and see that in both cases there is only one possibility, the neighborhood V7. We get:

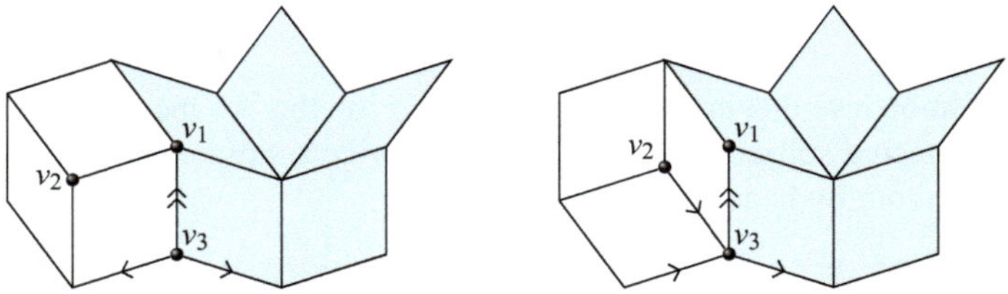

Next, we look at all possible neighborhoods of the vertex v_3 and see that the only possibility is V7 on the left and V4 on the right:

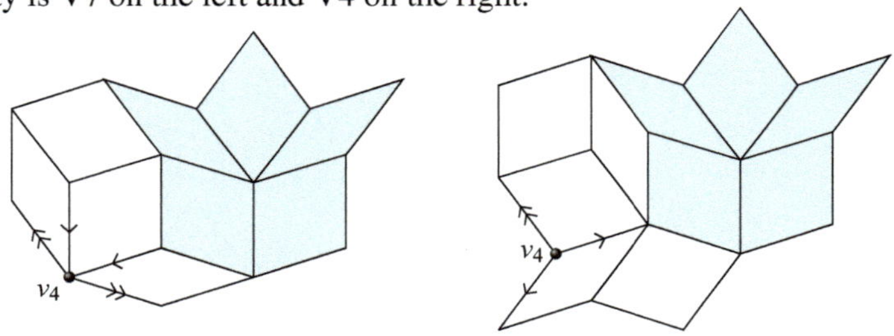

The vertex neighborhood of v_4 on the left can be number four or six, on the right must be V7. We get the following three partial tilings:

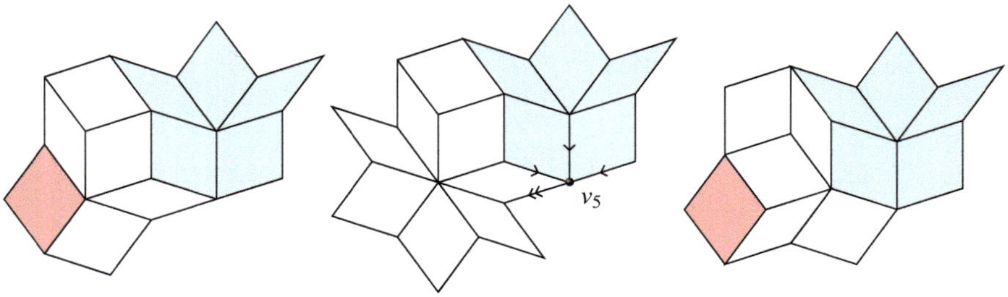

In red we see the forced tile. The first and third partial tilings are legal: one can check that they appear for example in the Cartwheel (Fig. 4.8). This proves that the dashed part in (4.10) is not forced by the blue patch.

In the picture in the middle the forced tile is not present, but such a partial tiling is illegal. Indeed, the vertex neighborhood of the vertex v_5 must be V5 or V6. In both cases the thin rhombus should have another thin rhombus attached along the free edge, which is impossible because it would overlap with a white tile:

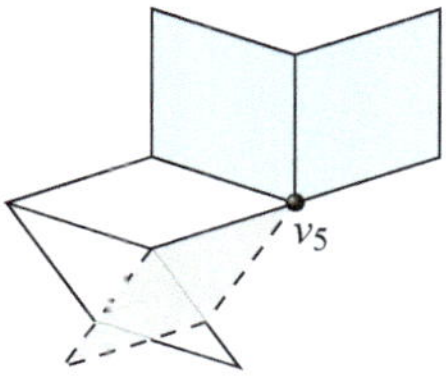

■

The example above illustrates an interesting feature of Penrose tilings: placing a tile can force the presence of other tiles "far away". On one side, this means that Penrose tilings exhibit a kind of non-local growth: if you place a tile on the plane, many other (forced) tiles may appear, even far from the patch you are extending. On the other side, no amount of local information is enough to ensure that the tiles we place will form a legal patch. This aspect of Penrose tilings is discussed in the seminal paper [Pen89].

4.6 The Three-Color Theorem

Let us recall some terminology from graph theory. A *directed graph* is a tuple (V, E, s, t) where V is a set whose elements we call *vertices*, E a set whose elements we call *edges*, $s, t : E \to V$ two maps called the *source* and the *target* map. Pictorially we draw a point for each vertex, an arrow from x to y for each edge with source x and target y (if any). For example

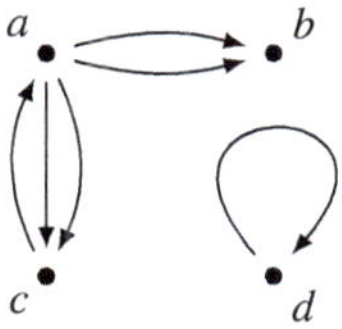

is a (finite) graph with vertex set $V = \{a, b, c, d\}$. An edge e with $s(e) = t(e)$ is called a *loop* (the arrow $d \to d$ in the above picture, for example).

We are interested in graphs with no parallel arrows, i.e. such that $s \times t : E \to V \times V$ is injective. These can also be defined as pairs (E, V) with $E \subseteq V \times V$ (we identify an edge with its ordered pair of endpoints).

One passes to *undirected* graphs by "forgetting" the arrow. More precisely, an undirected graph is a pair (V, E) where $E \subseteq V \times V$ is symmetric, i.e. for all $x, y \in V$, if $(x, y) \in E$ then $(y, x) \in E$. A *subgraph* of (V, E) is simply a graph (V', E') such that $V' \subseteq V$ and $E' \subseteq E$.

To an edge-to-edge (partial) tiling $\mathcal{T}$ we can associate a graph (undirected with no loops) as follows: take $V = \mathcal{T}$ and E the set of pairs $(T_1, T_2) \in \mathcal{T} \times \mathcal{T}$, with $T_1 \neq T_2$, such that T_1 and T_2 share an edge. Thus, vertices are tiles and edges are pairs of adjacent tiles. Recall from the proof of the Extension Theorem 2.14 that, if one has finitely many compact prototiles, then the tiles can be numbered. This means that

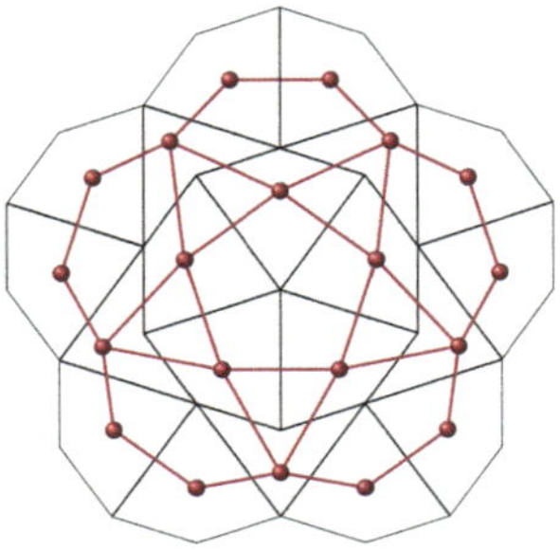

Fig. 4.11 A tiling by kites and darts and the associated undirected graph

Fig. 4.12 3-coloring of a tiling by kites and darts

we have a countable number of vertices and edges in the associated graph. Pictorially, given a tiling one can draw a point internal to each tile, and connect all pairs of points in adjacent tiles by a curve crossing the common edge. In Fig. 4.11 we see a partial tiling by kites and darts and (a picture of) the associated graph.

Let $k \geq 1$. A *k-coloring* of a graph (V, E) is a map $c : V \to \{1, \ldots, k\}$ such that, for each $(x, y) \in E$, $c(x) \neq c(y)$. For the graph of an edge-to-edge tiling, if we replace the numbers $1, \ldots, k$ by k different colors, a coloring is the assignment of a color to each tile such that no adjacent tiles have the same color. In Fig. 4.12 we see a 3-coloring of a partial tiling by kites and darts.

The *chromatic number* of an undirected graph is the minimum value of k such that the graph can be k-colored.

The celebrated four color theorem says that every *planar* graph can be 4-colored, i.e. it has chromatic number 4.[2]

A main tool for computing the chromatic number of an infinite graph, such as the one associated to an edge-to-edge tiling of the plane, is the following theorem by N.G. de Bruijn and P. Erdős.

Theorem 4.26 ([dBE51]) *Let $k \geq 1$. A graph is k-colorable if and only if all its finite subgraphs are k-colorable.*

From this theorem, in particular, it follows that all Penrose tilings (with the same protoset) have the same chromatic number, since every patch of a tiling appears in every other tiling with the same protoset. To compute, for example, the chromatic number of tilings with kites and darts, it is sufficient to consider a single example, e.g. the Cartwheel. Unfortunately, such a computation is anything but easy.

---

[2] We will not recall the definition of planar graph here.

It is easy to realize that:

Remark 4.27 Every Robinson tiling can be 2-colored.

Indeed, one simply associates to all A-tiles one color and to all B-tiles another color (like in Fig. 3.6). This gives a 2-coloring, since in a Robinson tiling, two tiles of the same type never share an edge.

In the case of kites and darts and of Penrose rhombi, the chromatic number obviously cannot be 2. Any tiling of the plane by kites and darts contains an ace (in fact, infinitely many), and the ace is not 2-colorable because its graph is the complete graph with three vertices: 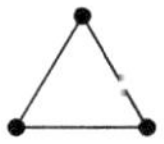 . Similarly, any tiling of the plane by Penrose rhombi contain infinitely many vertex neighborhoods V7 and V8, that are not 2-colorable for the very same reason.

It was proved in [SW00] that every tiling of the plane by Penrose rhombi has chromatic number 3. In fact, this is a special case of a more general statement:

Theorem 4.28 ([SW00]) *Any parallelogram tiling can be 3-colored.*

The proof is very short and elegant, and we will reproduce it here.

Recall that the *valence* of a vertex v in a graph (V, E) is the cardinality of the set $\{y \in V : (x, y) \in V\}$, i.e. the number of edges that have source x. A vertex y is *adjacent* to x if $(x, y) \in E$. If the graph is associated to a tiling, the valence of a vertex tells us how many tiles are adjacent to a given tile.

Lemma 4.29 *Let $k \geq 1$ and (V, E) be a graph (unoriented, with no loops). If every finite subgraph has a vertex with valence less than k, then (V, E) is k-colorable.*

Proof By Theorem 4.26, it is enough to show that every finite subgraph (V', E') of (V, E) is k-colorable. We do this by induction on the cardinality n of V' for $n \geq k$. If $n = k$, the statement is trivial. Assume by inductive hypothesis that, for a given $n \geq k$, every subgraph with n vertices is k-colorable. Let (V', E') be a subgraph with $n + 1$ vertices and $v_0 \in V'$ a vertex with valence less than k. By inductive hypothesis, there exists a k-coloring $c : V' \setminus \{v_0\} \to \{1, \ldots, k\}$ of the subgraph obtained from (V', E') by removing the vertex v_0 and all edges (x, y) with either $x = v_0$ or $y = v_0$. Since v_0 has at most $k - 1$ adjacent vertices, there is an integer $c_0 \in \{1, \ldots, k\}$ different from $c(y)$ for all y adjacent to v_0 in (V', E'). Assigning the value $c(v_0) := c_0$ we get a k-coloring of (V', E'). ■

In an edge-to-edge partial tiling, a tile that has at most two adjacent tiles is called an *elbow*. This means that the corresponding vertex in the associated graph has valence at most 2. As a special case of the previous lemma, if every patch of a tiling has an elbow, then the tiling can be 3-colored by induction: given a patch, find an elbow, color the remainder by induction, and then color the elbow with the free color, which must exist since it has at most two neighbors.

Theorem 4.28 now immediately follows from Lemma 4.29 and the next lemma.

Lemma 4.30 *Every patch of a parallelogram tiling has an elbow.*

Proof It is enough to prove the statement for a patch covering exactly a connected subset P of $\mathbb{R}^2$. In the general case, this will give an elbow for every connected component. Observe that P is a simple polygon with $n \geq 3$ edges.

By contradiction, suppose the patch has no elbow.
This means that each edge of P belongs to a different tile:

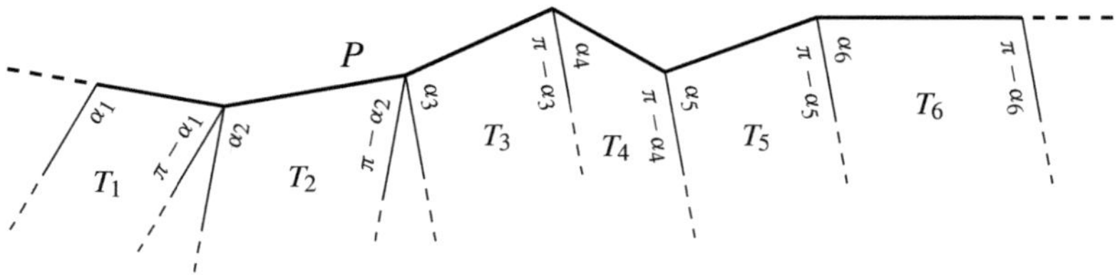

(There may be tiles with no external edges at all: in the picture we see two of them.) Each tile contributes π to the sum of internal angles of P, which would gives a total of $n\pi$, contradicting the well known fact that the sum of the internal angles of a simple n-gon is $(n-2)\pi$. ■

The same strategy unfortunately doesn't work for tilings by kites and darts. In Fig. 4.12 we see an example of a (3-colored) patch with no elbows.

A computer assisted proof that every tiling by kites and darts is 3-colorable was given in [Bab01].

Chapter 5
De Bruijn's Pentagrids

This chapter is devoted to the methods and ideas introduced by N.G. de Bruijn in [dB81], who gave some of the main inputs in the study of Penrose tilings. De Bruijn's methods concern tilings with Penrose rhombi:

To simplify the formulas, in the whole chapter we will assume that the edges of the rhombi have length 1, rather than φ like we did in Chap. 4. We also stress again that our single arrows have opposite orientation with respect to those in [dB81].

In the whole chapter:

$$\zeta := e^{2\pi \mathrm{i}/5}.$$

For future use, we define the following 5-tuples:

$$\vec{e}_j := \frac{1}{\sqrt{5}}(1, \zeta^j, \zeta^{2j}, \zeta^{3j}, \zeta^{4j}), \qquad j = -2, \dots, 2. \tag{5.1}$$

Because of (A.2), these vectors form an orthonormal basis of $\mathbb{C}^5$ equipped with the canonical scalar product, given by

$$\langle \vec{v}, \vec{w} \rangle := \sum\nolimits_{j=0}^{4} v_j^* w_j$$

for all $\vec{v} = (v_0, \dots, v_4)$ and $\vec{w} = (w_0, \dots, w_4)$ in $\mathbb{C}^5$. Any $\vec{x} \in \mathbb{C}^5$ can be uniquely written as $\vec{x} = \sum_{j=-2}^{2} \langle \vec{e}_j, \vec{x} \rangle \vec{e}_j$. For $\vec{x} \in \mathbb{R}^5$, this gives the decomposition

F. D'Andrea, *A Guide to Penrose Tilings*,
https://doi.org/10.1007/978-3-031-28428-1_5

$$x_j = \frac{1}{5}r + \frac{2}{5}\operatorname{Re}(z_1\zeta^{-j}) + \frac{2}{5}\operatorname{Re}(z_2\zeta^{-2j}), \tag{5.2}$$

where $r := \sum_{j=0}^{4} x_j$, $z_1 = \sum_{j=0}^{4} x_j\zeta^j$ and $z_2 = \sum_{j=0}^{4} x_j\zeta^{2j}$.

5.1 Pentagrids

In Sect. 4.4 we saw that ribbons, in some respects, behave similarly to lines and that every tiling by Penrose rhombi has associated five bundles of parallel ribbons pointing in five different directions. In fact, it is argued in [dB81] that, with a homeomorphism of the plane, one could transform the piecewise linear curves cutting ribbons in half—cf. (4.9)—into actual lines. This is what inspired the pentagrid method.

Given a unit vector $u \in \mathbb{C}$, a bundle of parallel lines, each at a fixed distance $\delta > 0$ from the next and orthogonal to u, will be called a *grid* with direction u. We will work with grids with step $\delta = 1$. The equation of such a grid, as a set of points in $\mathbb{C}$, is then:

$$\left\{z \in \mathbb{C} : \operatorname{Re}(zu^{-1}) + \gamma \in \mathbb{Z}\right\},$$

where $\gamma \in \mathbb{R}$ is a constant.

Points on the grid can be parametrized as:

$$z = (n + it - \gamma)u, \qquad n \in \mathbb{Z},\ t \in \mathbb{R}. \tag{5.3}$$

We will use the integer n to label lines on the grid, like in the following picture:

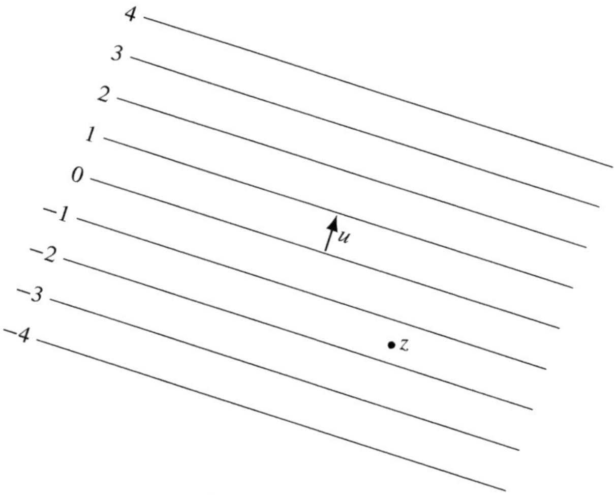

(5.4)

The grid defines a function $N : \mathbb{C} \to \mathbb{Z}$ given by

$$N(z) := \left\lceil \operatorname{Re}(zu^{-1}) + \gamma \right\rceil,$$

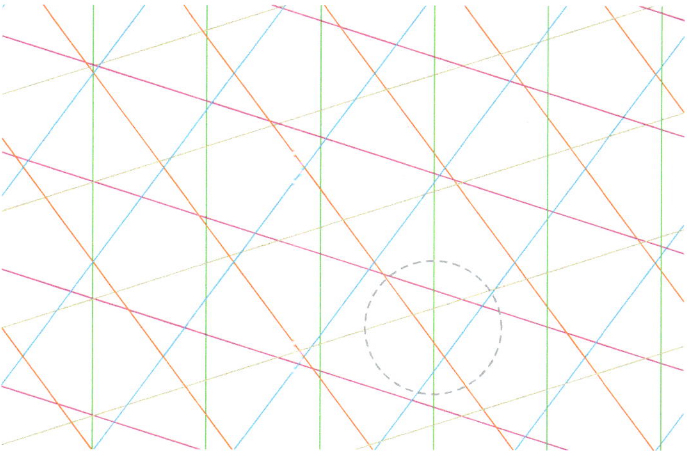

Fig. 5.1 A pentagrid. The grid with direction ζ^0 is green, the one with direction ζ is pink, the one with direction ζ^2 is blue, the one with direction ζ^3 is orange, the one with direction ζ^4 is olive

Fig. 5.2 Meshes

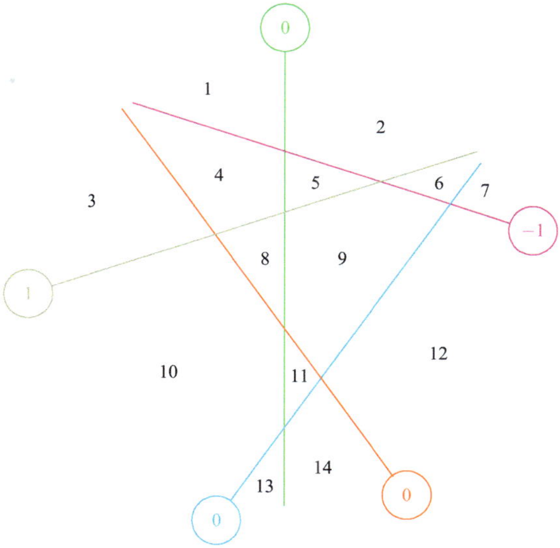

where $\lceil x \rceil := \min\{n \in \mathbb{Z} : n \geq x\}$ is the ceiling function. The point z in (5.4), for example, has $N(z) = -1$. The point z in (5.3) has $N(z) = n$.

The union of five grids with directions $(\zeta^j)_{j=0}^4$ and parameters $\vec{\gamma} = (\gamma_0, \ldots, \gamma_4)$ is called the *pentagrid* determined by $\vec{\gamma}$. An example is in Fig. 5.1.

Every pentagrid is the skeleton of an edge-to-edge tiling. A cell of this tiling will be called a *mesh*. In Fig. 5.2 one can see a zoom of the circled area of Fig. 5.1. Next to each line one can see its number. We also numbered the 14 meshes for future reference.

A pentagrid is *regular* if no more than two lines intersect at a point (thus, there are exactly four meshes incident at every vertex). Otherwise, it is called *singular*.

Each of the five grids of a pentagrid defines a function $N_j : \mathbb{C} \to \mathbb{Z}$ as explained before. The function $\vec{N} = (N_0, \ldots, N_4)$ is constant inside each mesh, and associates to each mesh five integer coordinates.

Below one can see the coordinates of the 14 meshes in Fig. 5.2:

$$\begin{array}{ll}
1 \to (0, 0, 1, 0, 1) & 8 \to (0, -1, 1, 0, 2) \\
2 \to (1, 0, 1, 0, 1) & 9 \to (1, -1, 1, 0, 2) \\
3 \to (0, -1, 1, 1, 1) & 10 \to (0, -1, 1, 1, 2) \\
4 \to (0, -1, 1, 0, 1) & 11 \to (1, -1, 1, 1, 2) \\
5 \to (1, -1, 1, 0, 1) & 12 \to (1, -1, 0, 0, 2) \\
6 \to (1, 0, 1, 0, 2) & 13 \to (0, -1, 0, 1, 2) \\
7 \to (1, 0, 0, 0, 2) & 14 \to (1, -1, 0, 1, 2)
\end{array}$$

To a mesh with coordinates $(n_0, \ldots, n_4)$ we associate the point

$$\sum_{j=0}^{4} n_j \zeta^j. \tag{5.5}$$

Two adjacent meshes have four out of five coordinates equal, and the remaining ones differ by 1. Every time we cross a line with orientation ζ^j, the coordinate n_j increases or decreases by 1, depending on the direction we move. The corresponding complex number (5.5) changes by $\pm\zeta^j$ every time we cross a line of the grid j.

In a regular pentagrid, each vertex is surrounded by four meshes, corresponding to four complex numbers that are the coordinates of the vertices of a rhombus with unit sides. For example:

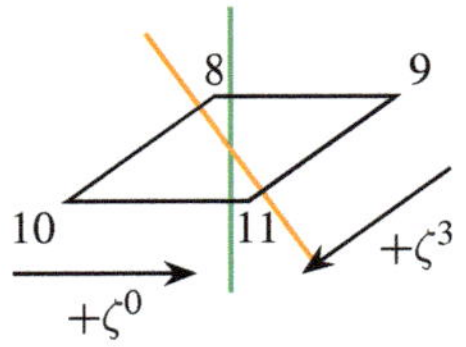

(5.6)

If we repeat the job for each mesh of a regular pentagrid, we get a tiling by Penrose rhombi to which we refer as the rhombus tiling defined by the pentagrid. In the next Fig. 5.3 we see the patch associated to the meshes in Fig. 5.2 and the arrows that are forced by the vertex neighborhoods. In the picture on the left, inside each ribbon we see the piecewise linear curve that roughly corresponds to a line of the same color in the pentagrid. Let us stress that, in general, the point (5.5) associated to a mesh is not inside the mesh itself and that a vertex of the pentagrid is not inside the rhombus generated by the surrounding meshes (the colored lines drawn in (5.6) are in a wrong position, and we put them there for visual purposes only).

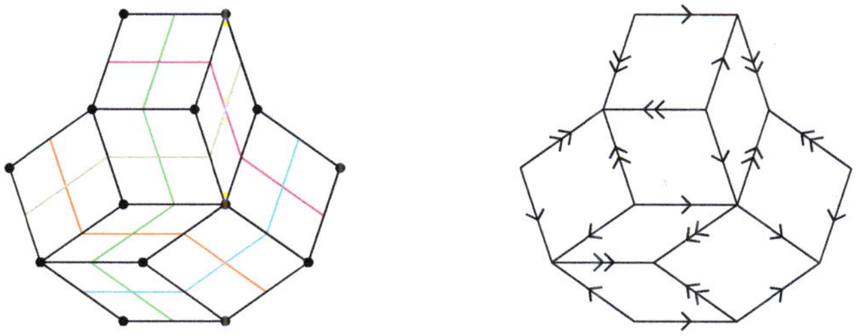

Fig. 5.3 The partial tiling constructed from the portion of pentagrid in Fig. 5.2

Proposition 5.1 *The rhombi associated with a regular pentagrid cover the plane.*

Proof Consider the map

$$f(z) := \sum_{j=0}^{4} N_j(z)\zeta^j.$$

The vertex set of the rhombus tiling is the image of f. When z runs along a curve, $f(z)$ draws points (vertices of rhombi) on the plane. When the curve crosses a line of the pentagrid, in the image of f we get two points joined by an edge of the tiling:

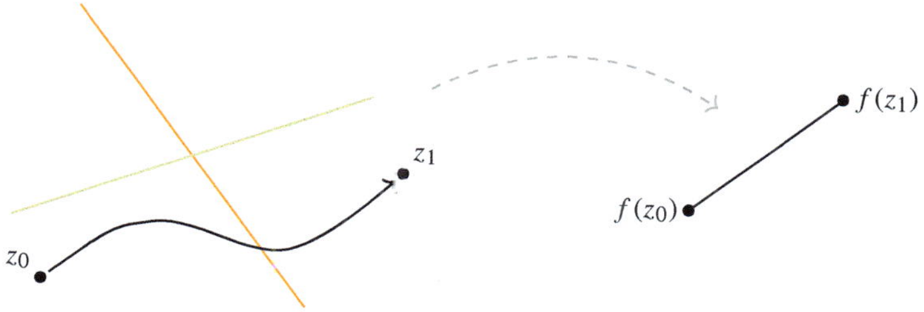

The only exception is when the curve crosses exactly a vertex of the pentagrid, but we can slightly modify it so that it does not intersect any vertex:

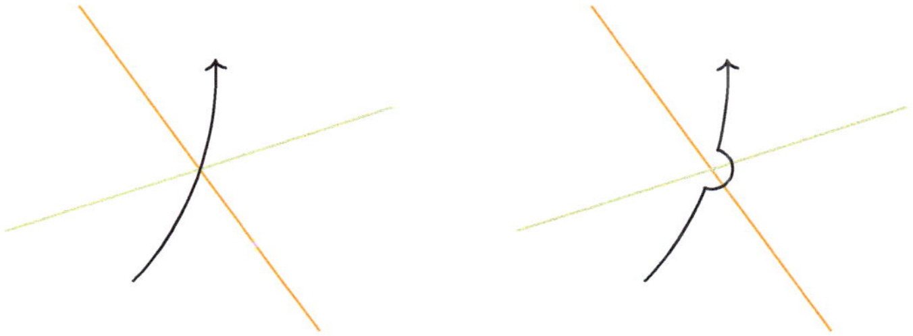

In this way, we transform curves into polygonal chains of the rhombus tiling.

Now, call

$$\lambda_j(z) := N_j(z) - \mathrm{Re}(z\zeta^{-j}) - \gamma_j. \tag{5.7}$$

By definition of N_j, one has

$$0 \le \lambda_j(z) < 1. \tag{5.8}$$

From (A.3) we deduce that

$$\sum_{j=0}^{4} \mathrm{Re}(z\zeta^{-j})\zeta^j = \sum_{j=0}^{4} \frac{z\zeta^{-j} + \bar{z}\zeta^j}{2}\zeta^j = \frac{5}{2}z,$$

which implies that

$$f(z) := \frac{5}{2}z + \sum_{j=0}^{4} (\lambda_j(z) + \gamma_j)\zeta^j.$$

It follows from the previous equation and (5.8) that $\left|f(z) - \frac{5}{2}z\right|$ is bounded by some constant C. Therefore, when z runs in a big circle of radius $r > \frac{2}{5}C$, in the image of f we get a polygon P contained in a disk D' of radius $\frac{5}{2}r + C$ and containing a disk D'' of radius $\frac{5}{2}r - C$ (see Fig. 5.4).

Now, we prove that D'' is covered by the tiles. When $r \to \infty$, this proves that we have rhombus tiling of the plane.

Take any point $z_0 \in D''$ and assume, by contradiction, that z_0 is not contained in a tile. Moving up at some point we will meet a segment of the polygon P (two, if we arrive at a vertex). This segment is the edge of a tile. Thus, moving up at some point we will cross a tile T:

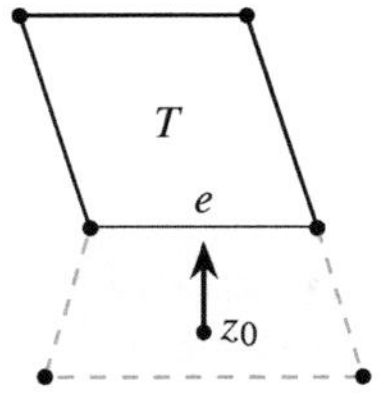

By hypothesis, the bottom edge e of T has no tile attached below. In terms of pentagrid, this means that T is a point of intersection a vertical line ℓ with some line ℓ' of the pentagrid forming an angle $\theta \in \{\pi/5, 2\pi/5\}$. Moving down, at some point we will meet another point of intersection (at most at a distance $1/\sin\theta$) where ℓ is crossed by another line from the same grid of ℓ':

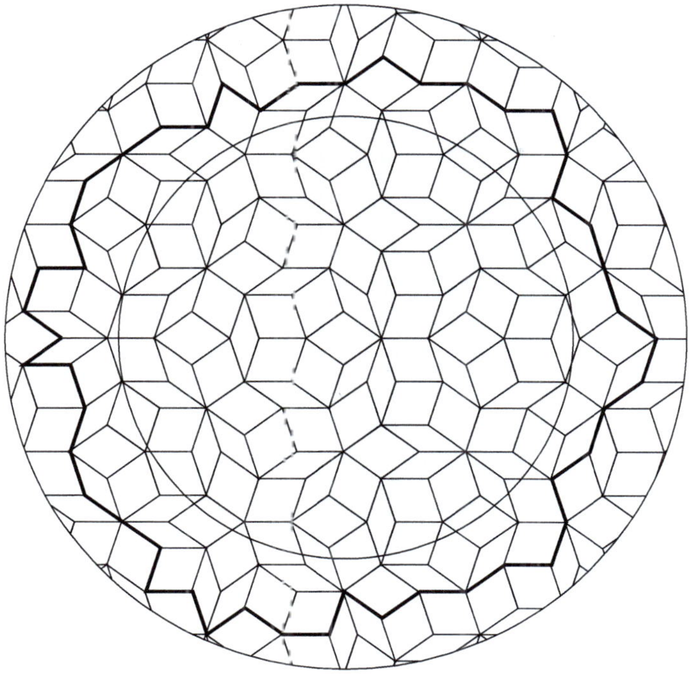

Fig. 5.4 When z runs in a big circle, $f(z)$ describes a big polygon

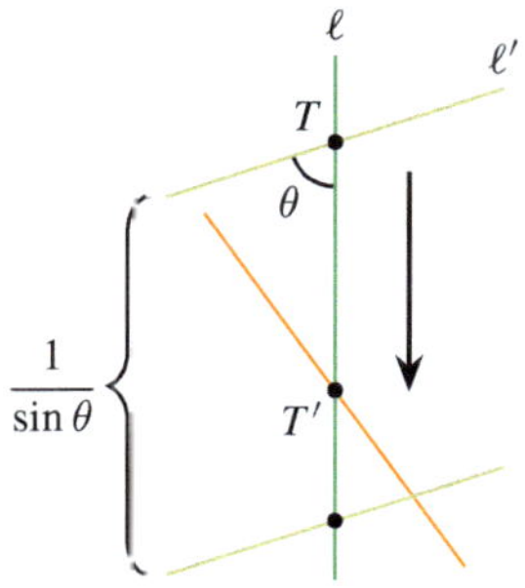

This point of intersection corresponds to a tile T' sharing the edge e with T, thus contradicting the initial assumption. ■

In the rest of this section we will assume that we have a regular pentagrid with parameter $\vec{\gamma}$ satisfying

$$\gamma_0 + \cdots + \gamma_4 = 0. \tag{5.9}$$

(In [dB81] this condition is part of the definition of a pentagrid.)

Remark 5.2 Without loss of generality, we can always assume that $\gamma_0 = 0$. Indeed, with a translation $z \mapsto z - \gamma_0$, we can transform a pentagrid with a general parameter $\vec{\gamma}$ satisfying (5.9) into one (defining an equivalent tiling) with parameter $\vec{\gamma}'$ given by $\gamma_j' = \gamma_j - \gamma_0 \operatorname{Re}(\zeta^j)$. In particular, $\gamma_0' = 0$. Furthermore, it follows from (A.3) that $\sum_{j=0}^4 \gamma_j' = \sum_{j=0}^4 \gamma_j$, so that the translation does not affect the condition (5.9).

Proposition 5.3 *In a regular pentagrid satisfying* (5.9)*:*

(i) The coordinates $(n_0, \dots, n_4) \in \mathbb{Z}^5$ *of a mesh satisfy:*

$$n_0 + n_1 + \cdots + n_4 \in \{1, 2, 3, 4\}. \tag{5.10}$$

(ii) Different meshes define distinct vertices of the tiling.

Thus, the map from meshes to vertices is bijective and it make sense to talk about <u>the</u> coordinates $(n_0, \dots, n_4)$ of a vertex (they are uniquely determined by the vertex).

Proof (i) Let λ_j be the functions in (5.7). Note that $\lambda_j(z) = 0$ if and only if z belongs to the grid j. Since a point belongs to at most two lines of a regular pentagrid, for every z at least three $\lambda_j(z)$ are not zero. Thus

$$0 < \sum_{j=0}^{4} \lambda_j(z) < 5 \ ,$$

where the second inequality comes from (5.8).

From (A.3) it follows that $\sum_{j=0}^4 \operatorname{Re}(z\zeta^j) = \operatorname{Re}(z \sum_{j=0}^4 \zeta^j) = 0$, and from the definition (5.7) and (5.9) we get:

$$\sum_{j=0}^{4} N_j(z) = \sum_{j=0}^{4} \lambda_j(z). \tag{5.11}$$

Since the left hand side is an integer, we deduce that its value is in $\{1, 2, 3, 4\}$.

(ii) We want to prove that the map

$$\mathbb{Z}^5 \to \mathbb{C}, \qquad (m_0, \dots, m_4) \mapsto \sum_{j=0}^{4} m_j \zeta^j \tag{5.12}$$

used in (5.5) becomes injective when restricted to coordinates of meshes of a pentagrid. This follows from (5.10).

Let $(n_0, \dots, n_4)$ and $(n_0', \dots, n_4')$ be the coordinates of two meshes defining the same vertex. Then $(m_0, \dots, m_4) := (n_0 - n_0', \dots, n_4 - n_4')$ is in the kernel of (5.12). This means that ζ is a root of the polynomial

$$m_0 + m_1 x + \cdots + m_4 x^4. \tag{5.13}$$

The 5th cyclotomic polynomial $1 + x + x^2 + x^3 + x^4$, which is the minimal polynomial of ζ over the field of the rational numbers, should then divide (5.13). Since they have the same degree, it must be $(m_0, \dots, m_4) = k(1, \dots, 1)$. Because of (5.10),

$$-3 \leq \sum_{j=0}^{4} m_j = 5k \leq 3,$$

which forces k to be 0. Hence $n_j = n'_j$ for all j. ■

The sum (5.10) is called the *index* of the vertex (5.5). The index increases by 1 if we move of one unit in the direction of ζ^j, and decreases by 1 if we move in the opposite direction, for all $j = 0, \dots 4$. It follows that, for both the thick and thin rhombus, there are only two possibilities (up to a rotation):

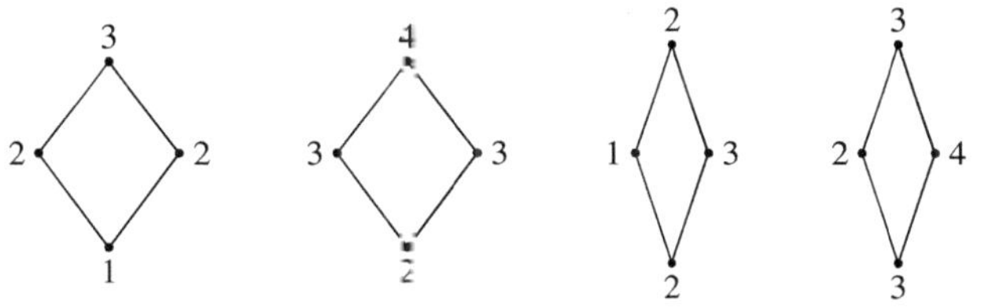

Lemma 5.4 *For a pentagrid to be regular, $\gamma_1 + \gamma_4 - \varphi^{-1}\gamma_0$ cannot be integer.*

Proof Let $\gamma_1 + \gamma_4 - \varphi^{-1}\gamma_0 \in 2\mathbb{Z} + \varepsilon$ with $\varepsilon \in \{0, 1\}$. Then, a triple point of intersection of the grids 0, 1 and 4 is given by $z = -\gamma_0 + \mathrm{i}(\gamma_1 - \gamma_4 + \varepsilon)/\sqrt{\varphi + 2}$. ■

Lemma 5.5 *Consider two rhombi in our tiling sharing an edge whose endpoints have indices 2 and 3. Let α and β be the internal angles of the rhombi at one of these two endpoints:*

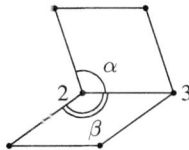

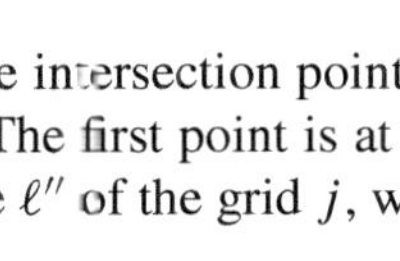

Then, α and β are either both acute or both obtuse.

Proof Consider two consecutive intersection points p and q on a line ℓ of the grid 0 (the other cases are similar). The first point is at the intersection with a line ℓ' of the grid i, the second with a line ℓ'' of the grid j, where $i, j \in \{1, 2, 3, 4\}$.

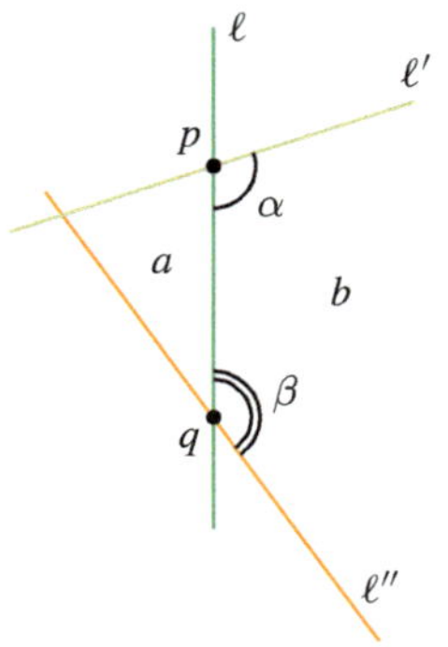

(5.13)

The segment $\overline{pq}$ corresponds, in the rhombus tiling, to an edge joining a vertex a to a vertex b (named like the corresponding meshes in (5.13)), with indices 2 and 3.

If we draw the picture (5.13) for all possible values of i and j such that $i + j$ is odd, we see that either both angles α and β are acute or they are both obtuse:

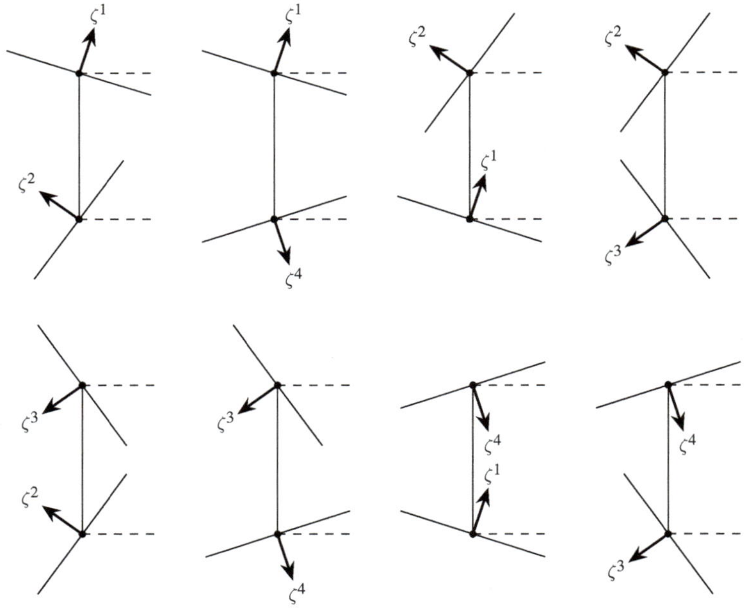

Now we shall show that $i + j$ is indeed odd, thus concluding the proof.

Without loss of generality, we can assume that $\gamma_0 = 0$ and ℓ is the imaginary axis (we can achieve this with a coordinate change, that does not affect the tiling, see also Remark 5.2). For a point $it \in \ell$ we have:

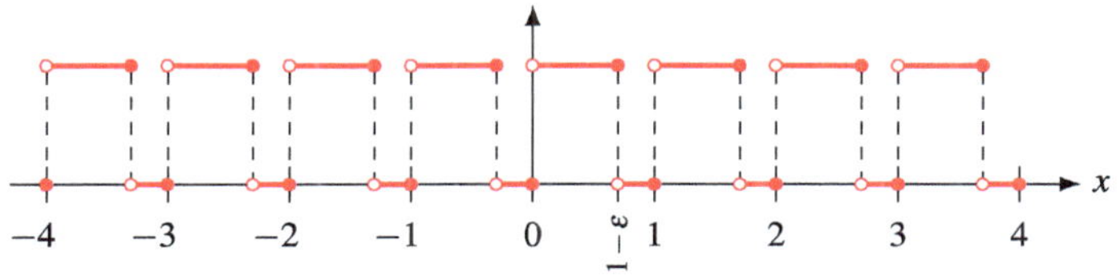

Fig. 5.5 Plot of the function (5.15)

$$N_1(it) = \left\lceil t \sin \tfrac{2\pi}{5} + \gamma_1 \right\rceil, \qquad N_3(it) = \left\lceil -t \sin \tfrac{4\pi}{5} + \gamma_3 \right\rceil,$$
$$N_2(it) = \left\lceil t \sin \tfrac{4\pi}{5} + \gamma_2 \right\rceil, \qquad N_4(it) = \left\lceil -t \sin \tfrac{2\pi}{5} + \gamma_4 \right\rceil.$$

Observe that

$$\lceil x \rceil + \lceil -x \rceil = \begin{cases} 0 & \text{if } x \in \mathbb{Z}, \\ 1 & \text{if } x \in \mathbb{R} \smallsetminus \mathbb{Z}. \end{cases}$$

From Lemma 5.4 we deduce that $\gamma_1 - \gamma_4 \notin \mathbb{Z}$, and since $\gamma_2 + \gamma_3 = -\gamma_1 - \gamma_4$, one has

$$\lceil \gamma_1 + \gamma_4 \rceil + \lceil \gamma_2 + \gamma_3 \rceil = 1. \tag{5.14}$$

Now, write $\gamma_1 + \gamma_4 = n - \varepsilon$ with $n \in \mathbb{Z}$ and $0 < \varepsilon < 1$. Then

$$N_1(it) + N_4(it) - \lceil \gamma_1 + \gamma_4 \rceil = \lceil x \rceil + \lceil -x - \varepsilon \rceil, \tag{5.15}$$

where $x := t \sin \frac{2\pi}{5} + \gamma_1$. When t runs from $-\infty$ to $+\infty$, the function (5.15) jumps from 0 to 1 at points where x is integer and from 1 to 0 at points where $x + \varepsilon$ is integer (cf. Fig. 5.5).

Since $N_1(it)$ is monotonically increasing and $N_4(it)$ monotonically decreasing with t, the function (5.15) jumps from 0 to 1 at the points where N_1 changes value, i.e. where the line ℓ intersects a line of the 1st grid, and it jumps from 1 to 0 at the points where N_4 changes value, i.e. where ℓ intersects a line of the 4th grid.

It follows from the argument above that the points of intersection with the 1st and 4th grid alternate, and studying the function

$$N_2(it) + N_3(it) - \lceil \gamma_2 + \gamma_3 \rceil \tag{5.16}$$

one proves that the same holds for the 2nd and 3rd grid. In particular, this means that it must be $i \neq j$.

Since both (5.15) and (5.16) take values in $\{0, 1\}$, from (5.14) we deduce that $N_1(it) + \cdots + N_4(it) \in \{1, 2, 3\}$.

Now, assume by contradiction that $i + j$ is even. Thus, either $\{i, j\} = \{1, 3\}$ or $\{i, j\} = \{2, 4\}$. Assume that moving along ℓ in the upward direction we meet first the intersection with the 1st grid, and then with the 3rd (thus $i = 3$ and $j = 1$). Since

(5.15) jumps from 0 to 1 at the intersection with the 1st grid, and (5.16) jumps from 1 to 0 at the intersection with the 3rd grid, we deduce that the sum

$$N_1(z) + \cdots + N_4(z) \tag{5.17}$$

is equal to 3 on the segment between p and q. For the very same reason, if $(i, j) = (1, 3)$ then (5.17) is 1 on the segment between p and q. Similarly, if $\{i, j\} = \{2, 4\}$ then (5.17) is either 1 or 3 on the segment between p and q.

But (5.17) is constant in the interior of $a \cup b$, while $N_0(z)$ is 0 in the mesh a and 1 in the mesh b. It follows that either the vertex a has index 1 and the vertex b has index 2, or a has index 3 and b has index 4, thus contradicting the assumption that one has index 2 and one has index 3. ■

We are now ready to prove the main result of this section: that there is a way (unique by Proposition 4.6) to put arrows on the edges of the rhombus tiling associated with a pentagrid, that gives a tiling by Penrose rhombi.

We will put arrows according to the following rule: double arrows go from a vertex with index 2 to a vertex with index 1, and from a vertex with index 3 to a vertex with index 4. Single arrows are put accordingly:

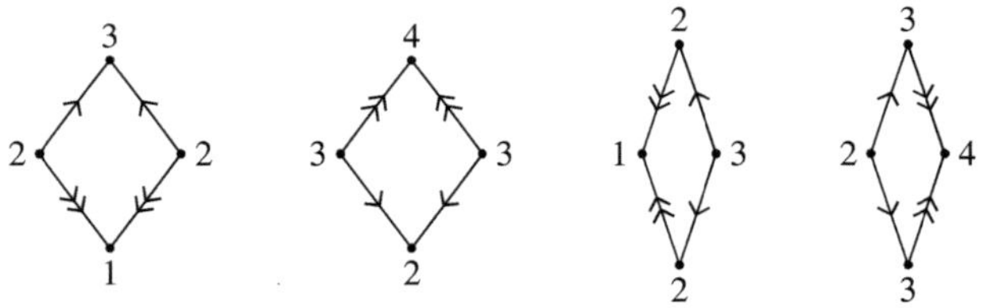

(5.18)

Proposition 5.6 *Given a regular pentagrid, with parameter $\vec{\gamma}$ satisfying* (5.9), *if we decorate the edges of the associated rhombus tiling as in (5.18), we get a tiling by Penrose rhombi (satisfying the matching rule).*

Proof By construction, all double arrows in the tiling will match. We must check that the matching condition for single arrows is satisfied as well. This is less obvious, since the orientation of the single arrow is not determined by the index of the end-points of the edge (sometimes is $2 \to 3$ and sometimes is $3 \to 2$). There are several configurations that would violate the matching rule, like for example:

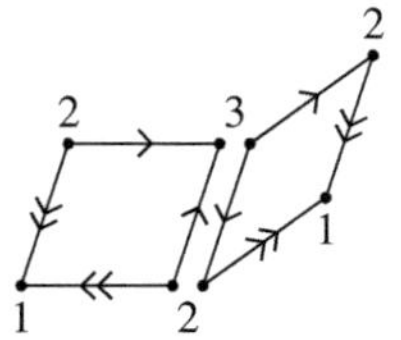

However, it immediately follows from Lemma 5.5 that these "wrong" configurations do not occur in our rhombus tilings. ■

5.2 The Cut-and-project Method

Using the pentagrid construction, we can now show that a tiling by Penrose rhombi is the 2-dimensional projection of a portion of a 5-dimensional lattice. It is useful to discuss the case of one-dimensional tilings first (these correspond to Conway worms in Penrose tilings, as explained in Sect. 4.3).

5.2.1 *One-dimensional Tilings*

In $\mathbb{R}^2$, consider the line ℓ with equation

$$x - \tau y = 0,$$

where τ is irrational (this guarantees that the only point of the lattice $\mathbb{Z}^2$ belonging to ℓ is the origin). We will assume that $\tau > 0$ (the cases $\tau < 0$ can be obtained from this with a reflection across the horizontal axis).

The orthogonal projection of a point (x, y) on the line ℓ is given by $\Pi(x, y)u$, where

$$u := \frac{1}{\sqrt{1+\tau^2}}(\tau, 1) \tag{5.19}$$

is a unit vector giving the direction of ℓ and

$$\Pi(x, y) := \frac{\tau x + y}{\sqrt{1+\tau^2}}.$$

We may think of $\Pi(x, y)$ as the coordinate of the projected point in the obvious one-dimensional reference frame on ℓ.

We now perform the following construction. If ℓ crosses the interior of a cell of the lattice $\mathbb{Z}^2$, then we project the top-left corner orthogonally on the line ℓ. The result is illustrated in Fig. 5.6, where the slope is given by the golden ratio.

In the picture we see: in gray the squares whose interior is crossed by ℓ, in red the points that are projected on ℓ, in light blue the polygonal chain (the *staircase*) connecting all these points. To the projected points we add the origin. We denote by Λ_τ the collection of red points (including the origin).

This construction gives us a partition of the line ℓ into segments of two different lengths: a white one with length $\tau/\sqrt{1+\tau^2}$, obtained by projecting the horizontal unit segments of the staircase, and a black one with length $1/\sqrt{1+\tau^2}$, obtained by projecting the vertical unit segments of the staircase.

We denote this one-dimensional tiling by $\mathcal{L}_\tau$.

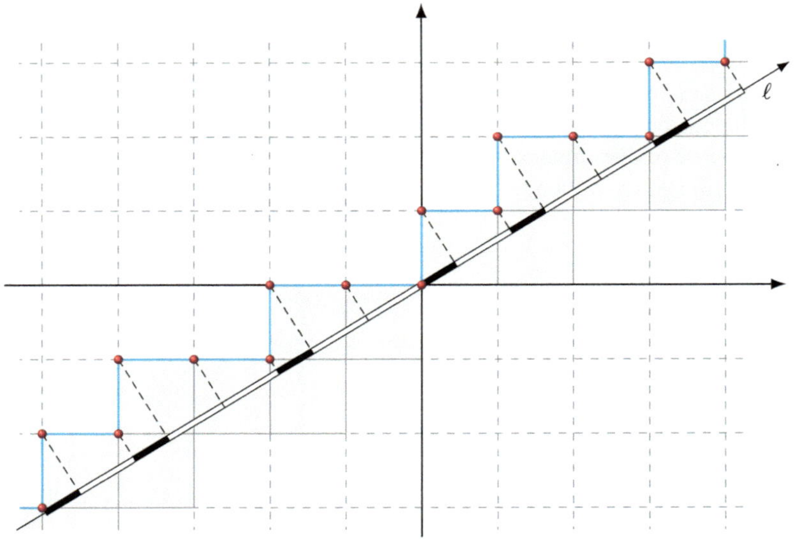

Fig. 5.6 The staircase projection

The line ℓ crosses the interior of a cell of the lattice $\mathbb{Z}^2$ if and only if it is below the top-left corner and at a distance less than δ_τ:

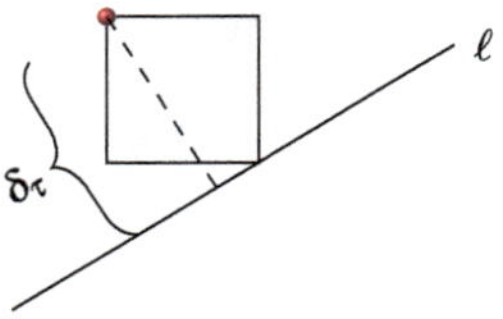

The constant δ_τ can be computed as the distance between ℓ and the point $(-1, 1)$. Using the point-line distance formula we get:

$$\delta_\tau = \frac{1+\tau}{\sqrt{1+\tau^2}}$$

Thus, Λ_τ is the intersection of $\mathbb{Z}^2$ with an half-open strip of width δ_τ (Fig. 5.7).

Proposition 5.7 *The subset $\Pi(\Lambda_\tau) \subset \mathbb{R}$ is non-periodic.*

Proof By contradiction, assume that $\Pi(\Lambda_\tau)$ is periodic with period $t_0 > 0$. The period is necessarily an integer combination of the lengths of the basic segments, that is

$$t_0 = \frac{m_0\tau + n_0}{\sqrt{1+\tau^2}} = \Pi(m_0, n_0),$$

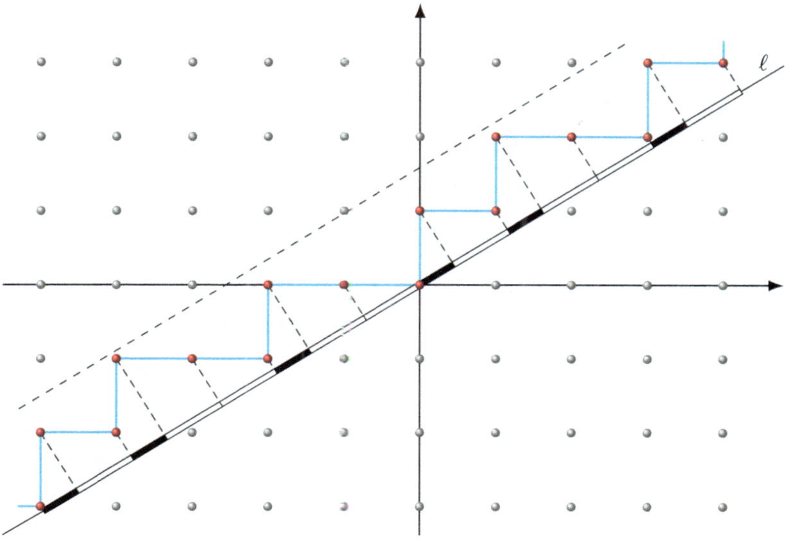

Fig. 5.7 Window of the projection

for some $m_0, n_0 \in \mathbb{Z}$. If $\Pi(\Lambda_\tau)$ is periodic, then kt_0 should belong to $\Pi(\Lambda_\tau)$ for all $k \in \mathbb{Z}$ (this is the orbit under translations of the origin). But

$$kt_0 = k\Pi(m_0, n_0) = \Pi(km_0, kn_0).$$

Since τ is irrational, the restriction of Π to $\mathbb{Z}^2$ is injective, and from $\Pi(km_0, kn_0) \in \Pi(\Lambda_\tau)$ we deduce that $p_k := (km_0, kn_0) \in \Lambda_\tau$ for all $k \in \mathbb{Z}$. The point p_k is at distance

$$|k|\frac{|m_0 - \tau n_0|}{\sqrt{1+\tau^2}}$$

from ℓ, and this is bigger than δ_τ for k big enough. We arrived at a contradiction. ∎

Recall that a one dimensional tiling with two segments, b and w, is called musical if every b is followed by a w one and there are no more than two consecutive w's in the tiling.

Proposition 5.8 *$\mathcal{L}_\tau$ is a musical tiling if and only if* $1 < \tau < 2$.

Proof ($\Leftarrow$) For the line ℓ to cross three squares, one on top of the other, it should cross both horizontal edges of the middle square, which would mean that its slope is $\tau^{-1} > 1$. For the line ℓ to cross four squares, one right to the other, it should cross three consecutive vertical edges, which would mean that $\tau^{-1} < 1/2$.

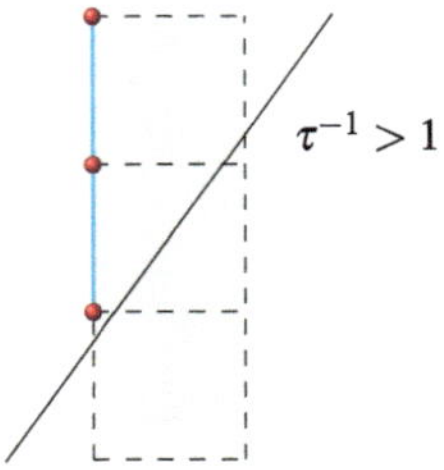

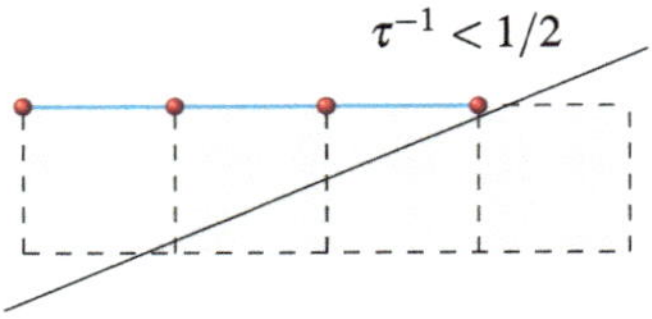

If $1/2 < \tau^{-1} < 1$, there are no consecutive vertical segments and at most two consecutive horizontal segments in the staircase, so that the black and white segments on ℓ satisfy the definition of musical tiling.

($\Rightarrow$) If $0 < \tau^{-1} < 1/2$, the three points $(0, 1)$, $(1, 1)$, $(2, 1)$ and $(3, 1)$ belong to Λ_τ. Thus we have a block www in our tiling, which means that it is not a musical one. Similarly if $\tau > 1$, the points $(0, 0)$, $(0, 1)$ and $(0, 2)$ belong to Λ_τ, we have a block bb, and the tiling is not a musical one. ∎

We now specialize the discussion to $\tau = \varphi$.

Proposition 5.9 *If we apply Composition to $\mathcal{L}_\varphi$ we obtain the tiling $-\varphi\mathcal{L}_\varphi$.*

Proof The projection Π maps corners

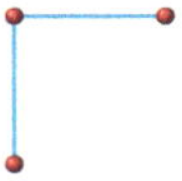

of the staircase to bw blocks, and the remaining horizontal unit segments of the staircase to w tiles. Composition transforms bw into a single segment, and to achieve this with a projection we have to replace each corner of the staircase by the diagonal of the corresponding square (cf. Fig. 5.8).

In doing so, we are removing from Λ_φ the lattice points that are above (and on) the line parallel to ℓ and crossing the point $(0, 1)$. This is the red line in Fig. 5.8, and it is at a distance

$$\delta' := \frac{\varphi}{\sqrt{1+\varphi^2}}$$

from ℓ (computed as usual with the point-line distance formula).

We call Λ' the set of points in Λ_φ that are at distance strictly less than δ' from ℓ.

It is easy to check that $\Lambda' \setminus \Lambda_\varphi$ contains all the corner points of the steps of the staircase, and only these points. The minimum distance of a corner point from ℓ is when the line crosses the point immediately below it:

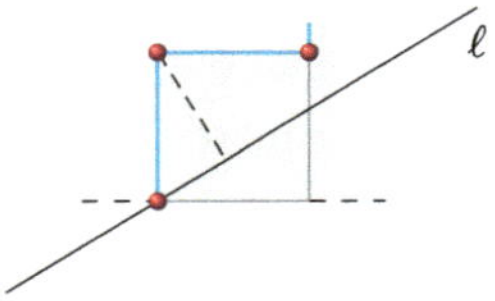

and is given exactly by δ'. This proves that no corner point is contained in Λ'. On the other hand, from the picture

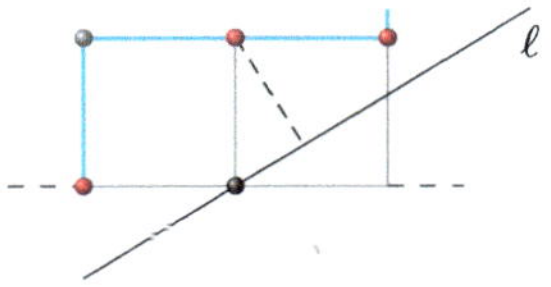

we see that every other point of Λ_φ has distance at most δ', with equality only when the line crosses the black point. This happens only if the black point is the origin (all the other points are strictly above ℓ), which gives the corner point $(0, 1)$. Thus, with a slight abuse of language, we proved that $\Pi(\Lambda')$ is the Composition of $\Pi(\Lambda_\varphi)$.

Now we want to transform the blue polygonal chain in Fig. 5.8 into a staircase. To do this, we apply the linear transformation represented (in the canonical basis) by the matrix

$$\varphi\begin{pmatrix} 0 & 1 \\ 1 & -1 \end{pmatrix}. \tag{5.20}$$

By diagonalizing the matrix one finds out that (5.19) is an eigenvector corresponding to the eigenvalue 1, and any non-zero orthogonal vector is an eigenvector with eigenvalue $-\varphi^2$. Thus, the matrix (5.20) represents a reflection across the line ℓ composed with a scaling of a factor φ^2.

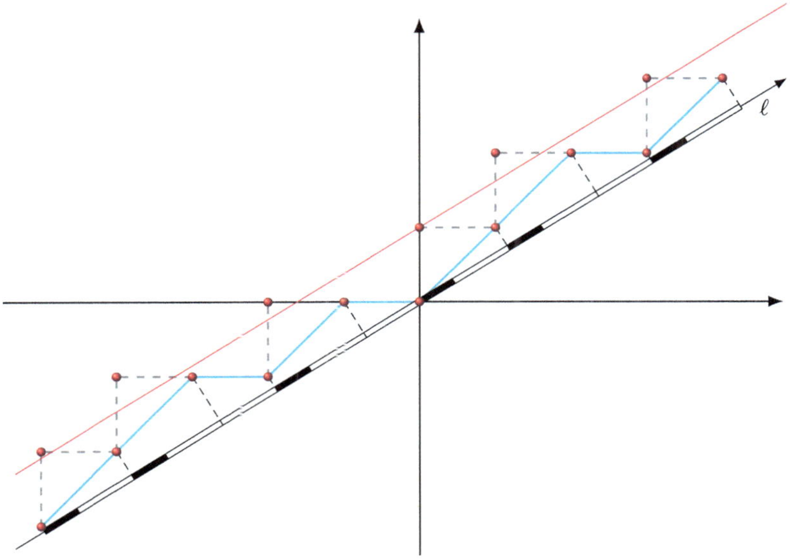

Fig. 5.8 Proof of Proposition 5.9

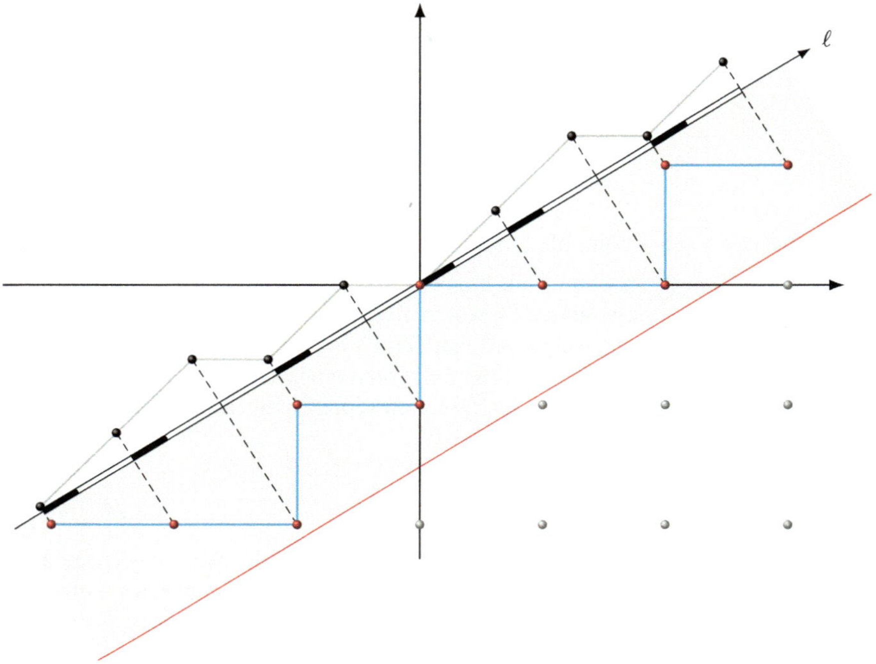

Fig. 5.9 Projection window for the composed tiling

This map does not change ℓ (its restriction to ℓ is the identity transformation). On the other hand, it transforms horizontal unit segments into vertical ones with length φ and the diagonal segments into horizontal ones with length φ:

The red line in Fig. 5.8 goes below ℓ and at a distance $\varphi^2\delta' = \varphi\delta_\varphi$. The situation is illustrated in Fig. 5.9. The tiling Composition of $\mathcal{L}_\varphi$ is obtained projecting the points of the lattice $\varphi\mathbb{Z}^2$ that are in the strip.

Now, if we apply a rotation of 180° the strip goes above ℓ, and if we rescale by a factor φ^{-1} what we obtain is the projection of points in the lattice $\mathbb{Z}^2$ above ℓ and at distance less than δ_φ. This gives exactly $\mathcal{L}_\varphi$. ■

Corollary 5.10 *$\mathcal{L}_\varphi$ is a Fibonacci tiling.*

Proof $\mathcal{L}_\varphi$ is a musical tiling, and if we apply Composition an arbitrary number of times we get a tiling that is similar to $\mathcal{L}_\varphi$. ■

One can obtain additional Fibonacci tilings replacing ℓ by the line with equation $x - \varphi y + c = 0$, with $c \in \mathbb{R}$. Then, we define $\Lambda_{\varphi,c}$ as the set of points of the lattice $\mathbb{Z}^2$ that are on the line or above the line, at a distance less than δ_φ, with δ_φ as before. If $c \in \mathbb{Z} + \varphi\mathbb{Z}$ the line contains exactly one point with integer coordinates, and if $c \notin \mathbb{Z} + \varphi\mathbb{Z}$ it contains none. Repeating the above construction we obtain a tiling $\mathcal{L}_{\varphi,c}$ depending on the constant c. The linear map (5.20) transforms the above-mentioned line into the line with equation $x - \varphi y + c' = 0$, with $c' = -c\varphi^2$, which becomes $x - \varphi y - c'\varphi^{-1} = 0$ after a 180° rotation and a scaling by φ^{-1}. Composition, then, transforms $\mathcal{L}_{\varphi,c}$ into $-\varphi\mathcal{L}_{\varphi,c\varphi}$.

5.2.2 Penrose Rhombi

In this section we explain how the tiling associated with a regular pentagrid, with parameter $\vec{\gamma} = (\gamma_0, \ldots, \gamma_4)$ satisfying (5.9), can be alternatively obtained by projecting lattice points from $\mathbb{R}^5$, in the same spirit we constructed musical tilings in Sect. 5.2.1 from the lattice $\mathbb{Z}^2$.

Consider the plane Π_γ in $\mathbb{R}^5$ with equations (in Cartesian coordinates):

$$\sum_{j=0}^{4} x_j = 0, \qquad \sum_{j=0}^{4} (x_j - \gamma_j)\zeta^{2j} = 0. \tag{5.21}$$

For $\vec{n} = (n_0, \ldots, n_4) \in \mathbb{Z}^5$, call (open unit hyper-)*cube* $\vec{n}$ the set:

$$\{x \in \mathbb{R}^5 : n_j - 1 < x_j < n_j \ \forall\, j = 0, \ldots, 4\},$$

and let

$$\Lambda_\gamma := \{\vec{n} \in \mathbb{Z}^5 : \Pi_\gamma \text{ intersects the cube } \vec{n}\}.$$

Proposition 5.11 *The vertices of the rhombus tiling, associated with a regular pentagrid with parameter* $\vec{\gamma} = (\gamma_0, \ldots, \gamma_4)$ *satisfying* (5.9)*, are the points* $n_0 + n_1\zeta + \cdots + n_4\zeta^4 \in \mathbb{C}$ *with* $\vec{n} \in \Lambda_\gamma$.

Proof In order to prove Proposition 5.11, we first rephrase the construction in more geometrical terms. Using the orthonormal basis (5.1) we rewrite (5.21) in the form $\langle \vec{e}_0, \vec{x} - \vec{\gamma}\rangle = \langle \vec{e}_{-2}, \vec{x} - \vec{\gamma}\rangle = 0$. Thus, $\vec{x} \in \mathbb{R}^5$ belongs to Π_γ if and only if $\vec{x} - \vec{\gamma}$ is orthogonal to the vectors $\vec{e}_0$, $\vec{e}_{-2}$ and $\vec{e}_2$ (being a real vector). That is,

$$\vec{x} = \vec{\gamma} + z\,\vec{e}_1 + z^*\vec{e}_{-1} \quad \text{for some } z \in \mathbb{C}.$$

Define two $\mathbb{R}$-linear maps by:

$$F : \mathbb{C} \to \mathbb{R}^5, \qquad z \mapsto z^* \vec{e}_1 + z\,\vec{e}_{-1},$$
$$G : \mathbb{R}^5 \to \mathbb{R}^5, \qquad \vec{v} \mapsto \langle \vec{e}_1, \vec{v} \rangle\, \vec{e}_1 + \langle \vec{e}_{-1}, \vec{v} \rangle\, \vec{e}_{-1}.$$

The map F is an isomorphism (of real vector spaces) onto its image, that we denote by V. The map G is the orthogonal projection from $\mathbb{R}^5$ to V. The composition

$$F^{-1} \circ G : \mathbb{R}^5 \longrightarrow \mathbb{C}$$

maps the canonical basis of $\mathbb{R}^5$ to the 5 vectors $(\zeta^j)_{j=0,\ldots,4}$, giving the direction of lines in a pentagrid. More generally,

$$n_0 + n_1\zeta + \cdots + n_4\zeta^4 = F^{-1} \circ G(\vec{n}), \qquad \forall\, \vec{n} \in \mathbb{Z}^5. \tag{5.22}$$

By the above discussion:

$$\Pi_\gamma = V + \vec{\gamma}.$$

The coordinates of a point $\vec{x} = F(z) + \vec{\gamma} \in \Pi_\gamma$ have then the form $x_j = \mathrm{Re}(z\zeta^{-j}) + \gamma_j$, for $j \in \{0, \ldots, 4\}$. The point belongs to the cube $\vec{n}$ if and only if

$$\lceil \mathrm{Re}(z\zeta^{-j}) + \gamma_j \rceil = n_j, \qquad \forall\, j = 0, \ldots, 4.$$

That is, z is a point in the pentagrid with $\vec{N}(z) = \vec{n}$, where $\vec{N} = (N_0, \ldots, N_4)$ are the ceiling functions of Sect. 5.1.

The same argument works the other way round. Regularity of the pentagrid guarantees that if $\vec{N}(z) = \vec{n}$, then $\mathrm{Re}(z\zeta^{-j}) + \gamma_j = n_j$ for at most two j's, and we can translate z a little bit in the two "problematic" directions in order to obtain a point in the (open) cube $\vec{n}$. ■

The map $F : \mathbb{C} \to V$ is an isometry in the sense of metric vector spaces, composed with a scaling of a factor $\sqrt{2}$. If we use F to identify the vector spaces $\mathbb{C}$ and V, we see from (5.22) that the vertices of our rhombus tiling are simply the orthogonal projections to V of lattice points in $\mathbb{R}^5$ contained in a suitable window. Note the similarity of this construction with that of Sect. 5.2.1 (except that in that case we project the top-left vertex of each square, rather than the top-right one).

We should stress that the pentagrid method gives rhombi with sides of length 1, while the projection from $\mathbb{R}^5$ to V gives rhombi of side $\sqrt{2}$.

While the projection of $\mathbb{Z}^2$ to a line with irrational slope is injective (cf. Sect. 5.2.1), the orthogonal projection from $\mathbb{Z}^5$ to V is not. But it becomes injective when we restrict the domain to Λ_γ (this is an immediate consequence of Proposition 5.3).

5.3 Composition and Pentagrids

In this section, we want to understand how the operation of Composition can be described in the language of pentagrids. We will adopt the notations of Sect. 5.1 and assume that we are given a regular pentagrid with parameter $\vec{\gamma}$ satisfying (5.9). We shall start with a characterization of the vertices in a rhombus tiling that, besides being used to prove the main theorem, has its own interest.

As a preparation, we now define two maps. Let $\vec{\gamma}$ satisfy (5.9). We define:

$$f_\gamma : \mathbb{Z}^5 \to \mathbb{Z} \times \mathbb{C}, \qquad f_\gamma(\vec{n}) := \Big(\sum_{j=0}^{4} n_j, \sum_{j=0}^{4} (n_j - \gamma_j)\zeta^{2j}\Big), \tag{5.23a}$$

$$g :]0, 1[^{\times 5} \to \mathbb{R} \times \mathbb{C}, \qquad g(\vec{\lambda}) := \Big(\sum_{j=0}^{4} \lambda_j, \sum_{j=0}^{4} \lambda_j \zeta^{2j}\Big). \tag{5.23b}$$

We also define four pentagons as follows: P_1 is the convex hull of the five roots of unity $1, \zeta, \zeta^2, \zeta^3, \zeta^4$ (its vertices), $P_2 := -\varphi P_1$, $P_3 := -P_2$ and $P_4 := -P_1$.

Lemma 5.12 *For all $r \in \{1, \ldots, 4\}$, one has*

$$\{z \in \mathbb{C} : (r, z) \in \mathrm{Im}(g)\} = \mathring{P}_r. \tag{5.24}$$

Proof Points of P_1 have the form $z = \sum_{j=0}^{4} \lambda_j \zeta^{2j}$, where $\sum_{j=0}^{4} \lambda_j = 1$ and $\lambda_j \geq 0$ for all j. If z is a vertex, then $\vec{\lambda}$ has four components equal to zero. If z is on an edge, then $\vec{\lambda}$ has at least three components equal to zero. Now we prove that every $z \in \mathring{P}_1$ can be written as above with all λ's different from zero, thus concluding the proof of (5.24) in the case $r = 1$.

If $z = 0$, then $z = \sum_{j=0}^{4} \zeta^{2j}/5$ and the proof is concluded. Assume that $z \neq 0$. In the next picture we see the pentagon P_1 and an internal point $z \neq 0$:

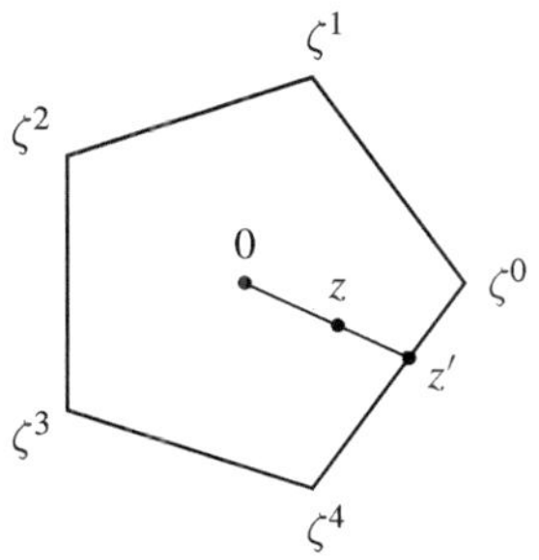

The point z' is at the intersection of the boundary with the line through 0 and z. Suppose that z' is on the segment from ζ^0 to ζ^4 like in the picture (the other cases are obtained by a rotation). Then, it can be written as a convex combination

$$z' = t'\zeta^0 + (1 - t')\zeta^4,$$

with $0 \leq t' \leq 1$.

The point z, on the other hand, is a convex combination of 0 and z':

$$z = t0 + (1-t)z' = t \sum_{j=0}^{4} \zeta^{2j}/5 + (1-t)t'\zeta^0 + (1-t)(1-t')\zeta^4$$

Here $0 < t < 1$, because by assumption z is not on the boundary and is not 0.

Simplifying, we get

$$z = \frac{t + 5(1-t)t'}{5}\zeta^0 + \frac{t}{5}\zeta^2 + \frac{t + 5(1-t)(1-t')}{5}\zeta^4 + \frac{t}{5}\zeta^6 + \frac{t}{5}\zeta^8 \tag{5.25}$$

and, since $0 < t < 1$, we see that all the coefficients are strictly greater than zero.

Note that the sum of any two coefficients in (5.25) is strictly less than one.

Next, we pass to the case $r = 2$. Let P_2' be the set of $\vec{\lambda} \in [0, 1]^{\times 5}$ with $\sum_{j=0}^{4} \lambda_j = 2$, and call P_2'' its image under the linear map $L : \vec{\lambda} \mapsto \sum_{j=0}^{4} \lambda_j \zeta^{2j}$. The image of a convex set is convex. Extreme points of P_2' are those $\vec{\lambda}$'s with two entries equal to 1 and the others equal to zero. Extreme points of P_2'' are of the form $L(\vec{\lambda})$ with $\vec{\lambda}$ an extreme point of P_2'. Thus, they are a subset of the points

$$\{\zeta^j + \zeta^{j+1}, \zeta^j + \zeta^{j+2}\}_{j=0,\dots,4}.$$

Points $\zeta^j + \zeta^{j+1} = \zeta^{j-2}(\zeta^2 + \zeta^3) = -\varphi\zeta^{j-2}$ are the vertices of the pentagon $P_2 = -\varphi P_1$. Points $\zeta^j + \zeta^{j+2}$ are not extreme, since

$$\zeta^j + \zeta^{j+2} = \tfrac{1}{\varphi+1}(-\varphi\zeta^j) + \tfrac{\varphi}{\varphi+1}(-\varphi\zeta^{j+3}).$$

Thus $P_2'' = P_2$, since they have the same extreme points. Now, we want to show that every internal point is the image of a point in the open cube.

If $z \in \mathring{P}_2 = -\varphi\mathring{P}_1$, we proved that $z = -\varphi \sum_{j=0}^{4} p_j\zeta^j$ for some $(p_0, \dots, p_4) \in]0, 1[^{\times 5}$ satisfying $\sum_{j=0}^{4} p_j = 1$. In fact, we can choose the coefficients so that $p_j + p_k < 1$ for all $j \neq k$. Therefore

$$z = \sum_{j=0}^{4} p_j(-\varphi\zeta^j) = \sum_{j=0}^{4} p_j(\zeta^j + \zeta^{j+1}) = \sum_{j=0}^{4}(p_j + p_{j-1})\zeta^j = \sum_{j=0}^{4} \lambda_j\zeta^j$$

where $p_{-1} := p_4$ and $\lambda_j := p_j + p_{j-1} \in]0, 1[$. Clearly $\sum_{j=0}^{4} \lambda_j = 2\sum_{j=0}^{4} p_j = 2$, hence the thesis.

Finally, consider the involution $\vec{\lambda} \mapsto \vec{\lambda}^\vee$ of the cube $]0, 1[^{\times 5}$ given by:

$$\vec{\lambda}^\vee := (1 - \lambda_0, \dots, 1 - \lambda_4).$$

Using (A.3), we deduce that

$$g(\vec{\lambda}) = (3, z) \iff g(\vec{\gamma}^\vee) = (2, -z) \quad \text{and} \quad g(\vec{\lambda}) = (4, z) \iff g(\vec{\gamma}^\vee) = (1, -z).$$

Thus, the cases $r = 3, 4$ follow from the cases $r = 2, 1$ respectively. ■

Lemma 5.13 *For all $z \in \mathbb{C}$ and all $a, b \in \mathbb{N}$ with $(a, b) \neq (0, 0)$, one has*

$$\sum_{j=0}^{4} \operatorname{Re}(z\zeta^{-aj})\zeta^{bj} = \frac{5}{2}\,\delta_{a,b}\,z. \tag{5.26}$$

Proof It immediately follows from $\sum_{j=0}^{4} \operatorname{Re}(z\zeta^{-aj})\zeta^{bj} = \frac{1}{2}\sum_{j=0}^{4}(z\zeta^{(b-a)j} + \bar{z}\zeta^{(b+a)j})$ using (A.3). ■

Proposition 5.14 *Consider a regular pentagrid with parameter $\vec{\gamma}$ satisfying* (5.9). *Then, $\sum_{j=0}^{4} n_j\zeta^j$ is a vertex of the associated rhombus tiling if and only if*

$$f_\gamma(\vec{n}) \in \operatorname{Im}(g), \tag{5.27}$$

where f_γ and g are the functions in (5.23).

Proof Note that

$$f_\gamma(\vec{n}) - g(\vec{\lambda}) = \sqrt{5}\Big(\langle \vec{n} - \vec{\gamma} - \vec{\lambda}, \vec{e}_0\rangle\,,\, \langle \vec{n} - \vec{\gamma} - \vec{\lambda}, \vec{e}_2\rangle \Big),$$

where $(\vec{e}_j)_{j=-2}^{2}$ is the orthonormal basis (5.1) and we used (5.9). Thus, (5.27) is equivalent to the condition that

$$\exists\, \vec{\lambda} \in]0, 1[^{\times 5}\colon \vec{n} - \vec{\gamma} - \vec{\lambda} \text{ is orthogonal to } \vec{e}_0, \vec{e}_2, \vec{e}_{-2} \tag{5.28}$$

(since a real vector orthogonal to $\vec{e}_2$ is orthogonal to $\vec{e}_{-2}$ as well).

For every $\vec{n} \in \mathbb{Z}^5$, we want to prove that (5.28) is equivalent to the existence of a mesh with coordinates $\vec{n}$, i.e. to the condition:

$$\exists\, z \in \mathbb{C} : n_j - 1 < \operatorname{Re}(z\zeta^{-j}) + \gamma_j < n_j \;\; \forall\, j = 0, \ldots, 4. \tag{5.29}$$

For the implication "(5.29) ⇒ (5.28)" simply define $\lambda_j := n_j - \operatorname{Re}(z\zeta^{-j}) - \gamma_j$ as in (5.7). From (5.26) it follows that $\langle \vec{n} - \vec{\gamma} - \vec{\lambda}, \vec{e}_0\rangle = \langle \vec{n} - \vec{\gamma} - \vec{\lambda}, \vec{e}_2\rangle = 0$.

In the other direction, (5.28) implies that $\vec{n} - \vec{\gamma} - \vec{\lambda}$ is a linear combination

$$\vec{n} - \vec{\gamma} - \vec{\lambda} = \frac{\sqrt{5}}{2}(z^*\vec{e}_1 + z\vec{e}_{-1}),$$

for some (unique) $z \in \mathbb{C}$. Thus, $n_j - \gamma_j - \lambda_j = \operatorname{Re}(ze^{-j})$ for all $j = 0, \ldots, 4$. Since now by hypothesis $0 < \lambda_j < 1$, the inequalities in (5.29) are satisfied. ■

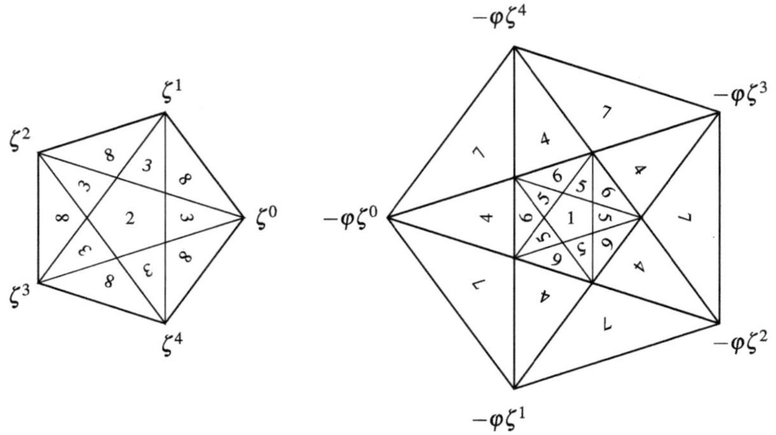

Fig. 5.10 The pentagon P_1 (on the left) and the pentagon $P_2 = -\varphi P_1$ (on the right)

Given a vertex of a tiling by Penrose rhombi, we shall say that it is of *type Vk*, with $k \in \{1, \ldots, 8\}$, if its vertex neighborhood is Vk (cf. Proposition 4.5).

Proposition 5.15 *Consider a regular pentagrid with parameter* $\vec{\gamma}$ *satisfying* (5.9). *The type of a vertex* $\sum_{j=0}^{4} n_j \zeta^j$ *of the associated tiling depends on the index* (5.10) *and on the complex number* $z_0 := \sum_{j=0}^{4} (n_j - \gamma_j)\zeta^{2j}$ *as follows:*

- *Index* 1*: we look at the pentagon* P_1 *in Fig. 5.10. The type is Vk if and only if* z_0 *lies in the open region marked by k.*
- *Index* 2*: we look at the pentagon* P_2 *in Fig. 5.10. The type is Vk if and only if* z_0 *lies in the open region marked by k.*
- *Index* 3*: similar to the case of index* 2*, but with* z_0 *replaced by* $-z_0$*.*
- *Index* 4*: similar to the case of index* 1*, but with* z_0 *replaced by* $-z_0$*.*

(If we compare Fig. 5.10 with Fig. 8 in [dB81], we see that there is a typo in [dB81]: the positions of the vertices K and Q—V3 and V8 in our notations—are exchanged.)

Proof With a slight abuse of terminology, we will identify a tuple $\vec{n} \in \mathbb{Z}^5$ with the corresponding vertex in the associated tiling. Let

$$\iota(\vec{n}) := \sum_{j=0}^{4} n_j \ , \qquad z_0(\vec{n}) := \sum_{j=0}^{4} (n_j - \gamma_j)\zeta^{2j},$$

be the components of the function (5.23b). The first one gives the index of the vertex $\vec{n}$, thus in particular has image in $\{1, \ldots, 4\}$. From Proposition 5.14 and (5.24) we deduce that

$$\vec{n} \text{ is a vertex with index } r \iff z_0(\vec{n}) \in \mathring{P}_r. \tag{5.30}$$

If $\vec{n}$ has index 3 (resp. 4), the vertex $\vec{n}^\vee := (1 - n_0, \ldots, 1 - n_4)$ has index 2 (resp. 1) and $z_0(\vec{n}^\vee) = -z_0(\vec{n})$ is in the interior of $P_3 = -P_2$ (resp. $P_4 = -P_1$). It is then enough to study the cases of index 1 and 2. In each of these two cases, we now study all possible sub-cases.

Let $(\vec{u}_j)_{j=0}^4$ be the canonical basis of $\mathbb{R}^5$, i.e. $\vec{u}_j$ has $(j+1)$th entry equal to 1 and every other entry equal to zero. A vertex is adjacent to that with coordinates $\vec{n}$ only if it has coordinates of the form $\vec{n} + s\vec{u}_j$, with $s \in \{\pm 1\}$ and $j \in \{0, \ldots, 4\}$ (but not all tuples of this form will give vertices of the tiling). Observe that

$$\iota(\vec{n} + s\vec{u}_j) = \iota(\vec{n}) + s \qquad \text{and} \qquad z_0(\vec{n} + s\vec{u}_j) = z_0(\vec{n}) + s\zeta^{2j}. \tag{5.31}$$

We will use repeatedly (5.31) together with (5.30) and (5.18), which tells us for each vertex neighborhood the possible indices of the vertices. For example, for V2 we have the following two possibilities:

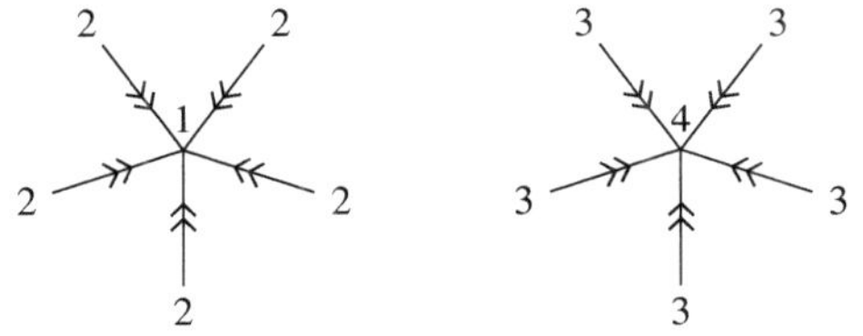

(5.32)

Index 1. We see from (5.18) that, for a vertex $\vec{n}$ with index 1, the neighborhood can only be V2, V3 or V8 (only ingoing double arrows).

If $\vec{n}$ has type V2, its neighboring vertices have index 2, cf. the picture on the left in (5.32): thus they are given by $\vec{n} + s\vec{u}_j$ with $s = +1$ and for all $j \in \{0, \ldots, 4\}$ (note that from this, using (5.5), we may deduce also the orientation of the vertex neighborhood). From the (5.30) we deduce that the type is V2 if and only if $z_0(\vec{n} + \vec{u}_j) = z_0(\vec{n}) + \zeta^{2j} \in \mathring{P}_2$ for all $j = 0, \ldots, 4$. That is, if and only if

$$z_0(\vec{n}) \in \bigcap_{j=0}^{4} (\mathring{P}_2 - \zeta^j). \tag{5.33}$$

This is the intersection of the five open pentagons in Fig. 5.11. Corners of $P_2 - \zeta^k$ have coordinates $-\varphi\zeta^i - \zeta^k$, $i, k \in \{0, \ldots, 4\}$. The dashed pentagon in Fig. 5.11 has corners

$$-\varphi\zeta^{j+2} - \zeta^j = -(1 + \zeta + \zeta^{-1})\zeta^{j+2} - \zeta^j = \zeta^{j+4} \qquad (j = 0, \ldots, 4),$$

i.e. it coincides with P_1. The intersection (5.33) is the shaded region in Fig. 5.11, corresponding to the region marked by 2 in Fig. 5.10. (With a little Euclidean geometry one also compute the coordinates if its corners, given by $-\varphi^{-2}\zeta^j$, with $j = 0, \ldots, 4$.)

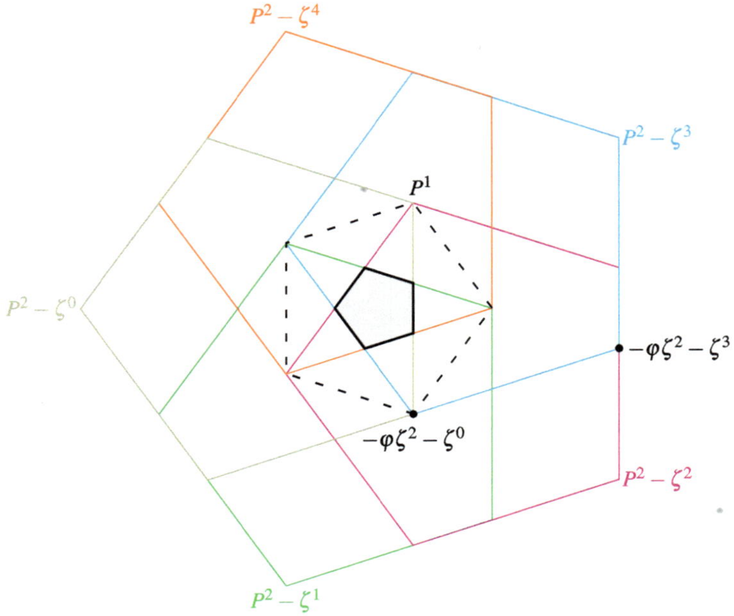

Fig. 5.11 The five pentagons $P_2 - \zeta^j$, $j = 0, \ldots, 4$

The next sub-cases are similar. The vertex type is V3 if and only if $\vec{n}$ has four neighboring vertices $\vec{n} + \vec{u}_j$. There are five possibilities: for each $k \in \{0, \ldots, 4\}$ we kill the vertex $\vec{n} + \vec{u}_k$, and this will give the position of the obtuse angle in V3. For example, if $k = 0$ we get the picture

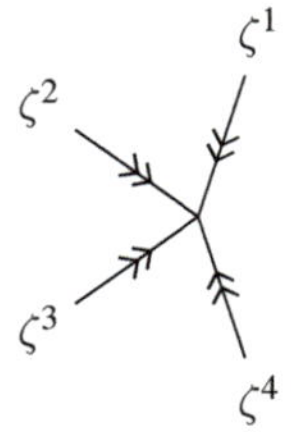

With a rotation we obtain the pictures for the other values of k.

By the same argument as above, the vertex type is V3 if and only if

$$z_0(\vec{n}) \in \bigcup_{k=0}^{4} \left(\left(\bigcap_{j \neq k} (\mathring{P}_2 - \zeta^j) \right) \setminus (\mathring{P}_2 - \zeta^k) \right),$$

i.e. if and only if $z_0(\vec{n})$ belongs to the open region in Fig. 5.11 that is the intersection of exactly four big pentagons. This coincides with the union of regions marked by

3 in Fig. 5.10. (Observe that each of these five regions corresponds to a different orientation of the vertex neighborhood.)

Finally, the type is V8 if and only if $z_0(\vec{n})$ is at the intersection of exactly three pentagons $\mathring{P}_2 - \zeta^j$. Combined with the fact that $z_0(\vec{n}) \in \mathring{P}_1$, this gives the remaining regions in Fig. 5.10.

Index 2. We see from (5.18) that, for a vertex $\vec{n}$ with index 2, the following cases are possible:

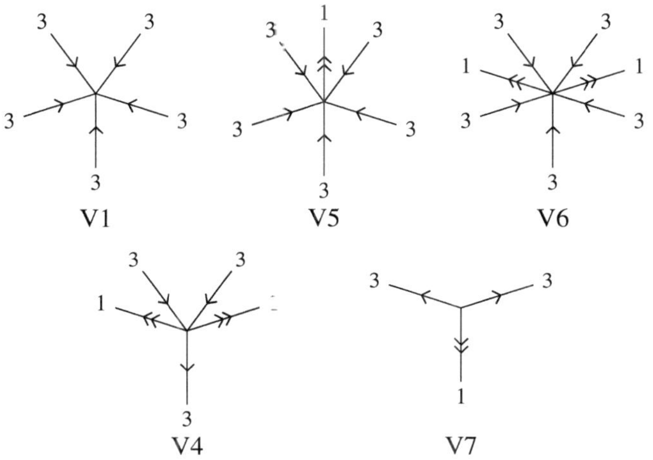

Every edge with a single arrow connects $\vec{n}$ with a vertex with index 3, hence of the form $\vec{n} + \vec{u}_j$; every edge with a double arrow connects $\vec{n}$ with a vertex with index 1, hence of the form $\vec{n} - \vec{u}_k$. The point $\vec{n} + \vec{u}_j$ is a vertex if and only if $z_0(\vec{n} + \vec{u}_j) \in -\mathring{P}_2$, i.e. $z_0(\vec{n}) \in -\mathring{P}_2 - \zeta^{2j}$; the point $\vec{n} - \vec{u}_k$ is a vertex if and only if $z_0(\vec{n} - \vec{u}_k) \in \mathring{P}_1$, i.e. $z_0(\vec{n}) \in \mathring{P}_1 + \zeta^{2k}$. Call $P_1^k := \mathring{P}_1 + \zeta^{2k}$ and $P_2^j := -\mathring{P}_2 - \zeta^{2j}$.

If $\vec{n}$ has type V1, V5 or V6, then $z_0(\vec{n})$ is at the intersection of the pentagons P_2^j, $j = 0, \ldots, 4$. This is the shaded region in Fig. 5.12. The dashed pentagon has vertices

$$\varphi\zeta^j - \zeta^{j+1} = -\varphi\zeta^{j+2},$$

with $j \in \{0, \ldots, 4\}$, i.e. it coincides with P_2. It follows that the shaded region coincides with the union of regions marked by 1, 5, 6 in Fig. 5.10. Consequently, the complement in P_2 corresponds to vertices of type V4 and V7. We distinguish between these two cases by counting the edges with a single arrow: V4 has three, meaning that corresponds to the case in which $z_0(\vec{n})$ is at the intersection of exactly three open pentagons in Fig. 5.12; the type V7 corresponds to the case when $z_0(\vec{n})$ is at the intersection of exactly two open pentagons, which combined with $z_0(\vec{n}) \in \mathring{P}_2$ gives the region marked by 7 in Fig. 5.12.

Finally, we focus on the shaded pentagon in Fig. 5.12, which coincides with $\varphi^{-1}P_1$ as one can easily check. In Fig. 5.13 we draw the five (open) pentagons P_1^k. The dashed pentagon is $\varphi^{-1}P_1$. The type of our vertex is V1 if and only if it has no

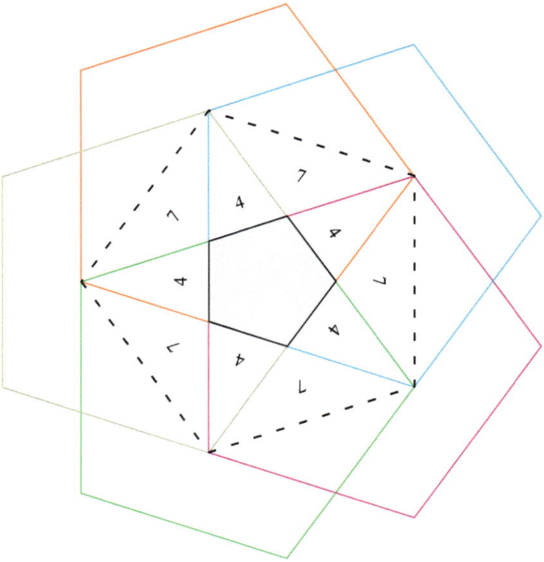

Fig. 5.12 The five pentagons $P_2 - \zeta^j$, $j = 0, \ldots, 4$

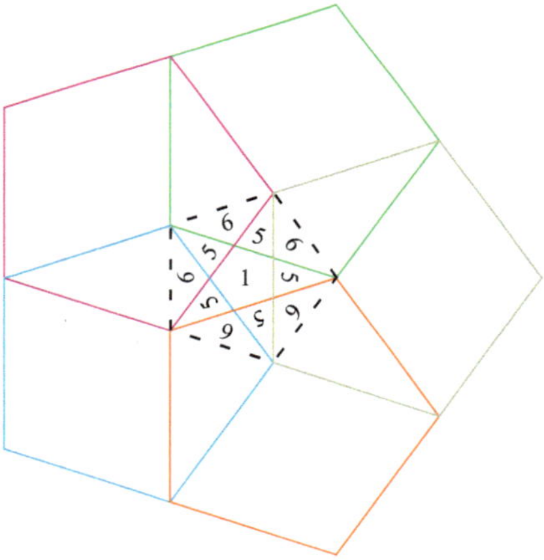

Fig. 5.13 The five pentagons $P_1 + \zeta^j$, $j = 0, \ldots, 4$

incident edges with double arrow, i.e. $z_0(\vec{n})$ does not belong to any of the pentagons P_1^k: this is the region marked by 1 in Fig. 5.13. The type is V5 if there is only one edge with a double arrow, i.e. $z_0(\vec{n})$ belongs to only one of the pentagons P_1^k: this is the region marked by 5 in Fig. 5.13. The remaining regions give V6. ■

We are now ready to prove the main theorem. The strategy is similar to the proof of Proposition 5.9. In that case, the effect of Composition was to remove the vertices projected from the corners of the staircase. In the present case, Composition removes the vertices of type V4 and V7, cf. Proposition 4.9, i.e. those corresponding to the outer regions of the pentagon P_2 in Fig. 5.10.

Theorem 5.16 *Let $\mathcal{T}_\gamma$ and $\mathcal{T}_{\gamma'}$ be two tilings associated with regular pentagrids with parameters $\vec{\gamma}$ and $\vec{\gamma}'$, respectively. Assume that the parameters satisfy* (5.9) *and that*

$$\gamma'_j = \gamma_{j-1} + \gamma_{j+1} \tag{5.34}$$

for all $0 \le j \le 4$ (here $\gamma_{-1} := \gamma_4$ and $\gamma_5 := \gamma_0$). Then, $\varphi\mathcal{T}_{\gamma'}$ is the Composition of $\mathcal{T}_\gamma$.

Proof We want to prove that the vertex set of $\varphi\mathcal{T}_{\gamma'}$ is given by all the vertices of $\mathcal{T}_\gamma$ of type different from V4 and V7. Let f_γ and g be as in (5.23), $\Lambda_\gamma \subset \mathbb{Z}^5$ the set of lattice points that are projected to vertices of $\mathcal{T}_\gamma$ (Proposition 5.11), and

$$W := \left\{\vec{n} \in \mathbb{Z}^5 : 1 \le \sum\nolimits_{j=0}^{4} n_j \le 4\right\}.$$

Recall that the projection Π from Λ_γ to vertices of $\mathcal{T}_\gamma$, given by $\Pi(\vec{n}) := \sum_{j=0}^{4} n_j \zeta^j$, is bijective (Proposition 5.3(ii)). With the same argument one proves that the restriction $f_\gamma|_W$ is injective. By Proposition 5.14, $\vec{n}$ belongs to Λ_γ if and only if $f_\gamma(\vec{n}) \in \mathrm{Im}(g)$. Since $\Lambda_\gamma \subset W$, the map f_γ gives a bijection from Λ_γ to $f_\gamma(W) \cap \mathrm{Im}(g)$.

For $\vec{\gamma}$ and $\vec{\gamma}'$ related by (5.34), we now construct a commutative diagram:

$$\begin{array}{ccc}
f_{\gamma'}(W) \cap \mathrm{Im}(g) & \xrightarrow{\;H\;} & f_\gamma(W) \cap \mathrm{Im}(g) \\
\uparrow {\scriptstyle f_{\gamma'}} & & \uparrow {\scriptstyle f_\gamma} \\
\Lambda_{\gamma'} & \xrightarrow{\;\Phi\;} & \Lambda_\gamma \\
\downarrow {\scriptstyle \Pi} & & \downarrow {\scriptstyle \Pi} \\
\text{vertices of } \mathcal{T}_{\gamma'} & \xrightarrow{\;\varphi\cdot\;} & \text{vertices of } \mathcal{T}_\gamma
\end{array} \tag{5.35}$$

All vertical arrows in (5.35) are bijections. The last horizontal map is the multiplication by φ in the complex plane. We must prove that it maps vertices of $\mathcal{T}_{\gamma'}$ to vertices of $\mathcal{T}_\gamma$, and in fact that gives a bijection with vertices of type different from V4 and V7, that is our thesis.

We initially define H and Φ on a bigger domain.

Firstly, $H : \{1, \ldots, 4\} \times \mathbb{C} \to \{1, \ldots, 4\} \times \mathbb{C}$ is the map

$$H(r', z'_0) := \left(3r' \bmod 5, -\varphi^{-1} z'_0\right),$$

where, with a slight abuse of notations, we denote by "j mod 5" the integer $j' \in \{1, \ldots, 4\}$ such that $j - j' \in 5\mathbb{Z}$. The first component of H is the permutation

$$\begin{array}{cccc} 1 & 2 & 3 & 4 \\ \downarrow & \downarrow & \downarrow & \downarrow \\ 3 & 1 & 4 & 2 \end{array}$$

and H is evidently bijective. It's restriction in (5.35) is then injective.

Next, we define a map $\Phi : W \to W$ by $\Phi(\vec{n}') = \vec{n}$ with:

$$n_j := \begin{cases} n'_{j-1} + n'_j + n'_{j+1} & \text{if } \imath(\vec{n}') = 1, \\ n'_{j-1} + n'_j + n'_{j+1} - 1 & \text{if } \imath(\vec{n}') = 2, \\ n'_{j-1} + n'_j + n'_{j+1} - 1 & \text{if } \imath(\vec{n}') = 3, \\ n'_{j-1} + n'_j + n'_{j+1} - 2 & \text{if } \imath(\vec{n}') = 4, \end{cases}$$

where $\imath(\vec{n}) = \sum_{j=0}^{4} n_j$ is the index. The map Φ is bijective (and then its restriction to $\Lambda_{\gamma'}$ is injective), with inverse given by

$$n'_j = \begin{cases} n_{j-1} + n_j & \text{if } \imath(\vec{n}) \in \{1, 2\}, \\ n_{j-1} + n_j - 1 & \text{if } \imath(\vec{n}) \in \{3, 4\}, \end{cases}$$

A simple computation gives

$$-\varphi^{-1} \sum_{j=0}^{4} (n'_j - \gamma'_j)\zeta^{2j} = \sum_{j=0}^{4} (n'_j - \gamma'_j)(1 + \zeta^2 + \zeta^{-2})\zeta^{2j} = \sum_{j=0}^{4} (n_j - \gamma''_j)\zeta^{2j},$$

where the sum on the subscript is mod 5 and $\gamma''_j := \gamma'_{j-1} + \gamma'_j + \gamma'_{j+1}$. Using (5.34) and (5.9) we find $\vec{\gamma}'' = \vec{\gamma}$, which proves that

$$H\big(f_{\gamma'}(\vec{n}')\big) = f_{\gamma}\big(\Phi(\vec{n}')\big), \tag{5.36}$$

for all $\vec{n}' \in W$. Thus, H maps $f_{\gamma'}(W)$ into $f_{\gamma}(W)$.

Next, for all $(r', z'_0) \in \operatorname{Im}(g)$ with $r' \in \{1, 2, 3, 4\}$ we prove that $(r, z_0) := H(r', z'_0)$ belongs to $\operatorname{Im}(g)$, so that the top horizontal map in (5.35) is well defined. In fact, from Fig. 5.10 we deduce more:

- if $r' = 1$, then $r = 3$ and $-z_0 \in \varphi^{-1}\mathring{P}_1$, meaning that it falls into one of the open regions in P_2 marked by 1, 5 or 6;
- if $r' = 2$, then $r = 1$ and $z_0 \in -\varphi^{-1}\mathring{P}_2 = \mathring{P}_1$,
- if $r' = 3$, then $r = 4$ and $-z_0 \in -\varphi^{-1}\mathring{P}_2 = \mathring{P}_1$,
- if $r' = 4$, then $r = 2$ and $z_0 \in \varphi^{-1}\mathring{P}_1$, meaning that it falls into one of the open regions in $\mathring{P}_2$ marked by 1, 5 or 6.

Thus, the function $\Pi \circ f_\gamma^{-1} \circ H$ maps its domain in (5.35) to vertices of $\mathcal{T}_\gamma$ of type different from V4 and V7. Every vertex of type different from V4 and V7 is in the image of this map, as one can check case by case: start with (r, z_0) in one of the above open regions, define $(r', z_0') := H^{-1}(r, z_0)$ (H^{-1} is well defined on $\{1, 2, 3, 4\} \times \mathbb{C}$), and check that $z_0' \in \mathring{P}_{r'}$.

It follows from (5.36) that the middle horizontal map is also well-defined – that is, Φ maps $\Lambda_{\gamma'}$ to Λ_γ – and that the top square in the diagram is commutative.

Finally, for all $\vec{n}' \in W$ and $\vec{n} = \Phi(\vec{n}')$:

$$\Pi(\vec{n}) = \sum_{j=0}^{4} n_j \zeta^j = \sum_{j=0}^{4} n_j' \zeta^j (1 + \zeta + \zeta^{-1}) = \varphi \sum_{j=0}^{4} n_j' \zeta^j = \varphi \cdot \Pi(\vec{n}'),$$

proving that the bottom horizontal map in (5.35) sends vertices of $\mathcal{T}_{\gamma'}$ (injectively) to vertices of $\mathcal{T}_\gamma$ of type different from V4 and V7, and that the bottom square in the diagram is commutative. ■

5.4 Congruence Classes of Tilings

A natural question is how many inequivalent tilings we can get from regular pentagrids. For starters we observe that, due to (5.9), in the construction of a pentagrid we have the freedom to choose four independent real parameters. However, the tiling coming from a pentagrid with parameters $\vec{\gamma}$ depends only on one (complex) parameter, given by

$$\omega_\gamma := \sum_{j=0}^{4} \gamma_j \zeta^{2j}. \tag{5.37}$$

This follows immediately from the cut-and-project method, and the fact that the second equation in (5.21) can be written in the form:

$$\sum_{j=0}^{4} x_j \zeta^{2j} = \omega_\gamma.$$

Thus, the plane Π_γ that we use to "cut" unit cubes in $\mathbb{R}^5$ (the only variable in the construction) depends only on ω_γ. We denote by Λ_γ, as in the previous sections, the set of points in $\mathbb{Z}^5$ that are projected to vertices of the tiling.

Let $\mathcal{I} \subset \mathbb{C}$ be the set of points (5.5) with index 0, that is

$$\mathcal{I} := \left\{ z = \sum\nolimits_{j=0}^{4} k_j \zeta^j \ : \ k_0, \ldots, k_4 \in \mathbb{Z} \text{ and } k_0 + \cdots + k_4 = 0 \right\}. \tag{5.38}$$

This is the ideal in $\mathbb{Z}[\zeta]$ generated by $1 - \zeta$.

Lemma 5.17 *The set $\mathcal{I}$ is dense in $\mathbb{C}$.*

Proof Recall that, if a and b are non-zero real numbers and a/b is irrational, then $a\mathbb{Z} + b\mathbb{Z}$ is dense in $\mathbb{R}$ (by Kronecker's approximation theorem). The set $\mathcal{I}$ is an abelian subgroup of $\mathbb{C}$ and contains the following four elements:

$$5 = 4\zeta^0 - \zeta^1 - \zeta^2 - \zeta^3 - \zeta^4, \qquad \mathrm{i}\sqrt{\varphi+2} = \zeta^1 - \zeta^4,$$
$$5\varphi = 2\zeta^0 + 2\zeta^1 - 3\zeta^2 - 3\zeta^3 + 2\zeta^4, \qquad \mathrm{i}\varphi^{-1}\sqrt{\varphi+2} = \zeta^2 - \zeta^3.$$

Since $5\mathbb{Z} + 5\varphi\mathbb{Z}$ is dense in $\mathbb{R}$, and $\mathrm{i}\sqrt{\varphi+2}(\mathbb{Z} + \varphi^{-1}\mathbb{Z})$ is dense in $\mathrm{i}\mathbb{R}$, the abelian group generated by the above four elements is dense in $\mathbb{C}$. ■

Proposition 5.18 *Let $\mathcal{T}$ and $\mathcal{T}'$ be tilings associated with regular pentagrids with parameters $\vec{\gamma}$ and $\vec{\gamma}'$ satisfying* (5.9). *Then, $\mathcal{T}$ and $\mathcal{T}'$ are*

(i) equal if and only if $\omega_\gamma = \omega_{\gamma'}$;
(ii) equivalent if and only if $\omega_\gamma - \zeta^{2k}\omega_{\gamma'} \in \mathcal{I}$ for some (unique) $k \in \{0, \ldots, 4\}$.

In the second case, writing

$$\omega_\gamma - \zeta^{2k}\omega_{\gamma'} = \sum_{j=0}^{4} m_j\zeta^{2j} \quad \textit{with} \sum_{j=0}^{4} m_j = 0 \tag{5.39}$$

one has $\mathcal{T} = \zeta^{-k}\mathcal{T}' + \sum_{j=0}^{4} m_j\zeta^j$.

Proof (i) Assume that $\omega_\gamma \neq \omega_{\gamma'}$. Let $\vec{n} \in \mathbb{Z}^5$ have $\sum_{j=0}^{4} n_j = r \in \{1, \ldots, 4\}$, that means $\sum_{j=0}^{4} n_j\zeta^{2j} = \sum_{j=0}^{4} m_j\zeta^j + r \in \mathcal{I} + r$, where $\vec{m} = (n_0 - r, n_3, n_1, n_4, n_2)$. It follows from Lemma 5.17 that $\mathcal{I} + r$ is dense in $\mathbb{C}$, hence there exists $\vec{n}$ such that

$$\sum_{j=0}^{4} n_j\zeta^{2j} - \omega_\gamma \in \mathring{P}_r \quad \text{and} \quad \sum_{j=0}^{4} n_j\zeta^{2j} - \omega_{\gamma'} \notin \mathring{P}_r,$$

where $P_1, \ldots, P_4$ are the pentagons of Sect. 5.3. From Proposition 5.15 we deduce that the point $\sum_{j=0}^{4} n_j\zeta^j$ belongs not $\mathcal{T}$ but not to $\mathcal{T}'$. Thus, $\mathcal{T} \neq \mathcal{T}'$.

(ii) Since the vertices of $\mathcal{T}$ and $\mathcal{T}'$ belong to the ring $\mathbb{Z}[\zeta]$, the two tilings are equivalent if and only if

$$\mathcal{T} = \zeta^{-k}\mathcal{T}' + \sum_{j=0}^{4} m_j\zeta^j \tag{5.40}$$

for some $k \in \{0, \ldots, 4\}$ and $\vec{m} \in \mathbb{Z}^5$. Call $r := \sum_{j=0}^{4} m_j$. Since $\vec{m}$ and $\vec{m} + (1, \ldots, 1)$ correspond to the same point in $\mathbb{C}$, we can always choose $\vec{m}$ so that $r \in \{0, \ldots, 4\}$.

The transformation $\sum_{j=0}^{4} n_j\zeta^j \mapsto \sum_{j=0}^{4}(n_j + m_j)\zeta^j$ on vertices of $\mathcal{T}'$ increases by r their index. Since the tiling $\mathcal{T}'$ has at least one vertex with index 4 (because every vertex neighborhood appears infinitely many times in a rhombus tiling and in every possible orientation), the tiling $\mathcal{T}$ has a vertex with index $r+4$. But the index cannot be greater than 4, therefore it must be $r=0$.

We now show that (5.40) is equivalent to (5.39). On one side, if (5.39) holds, then

$$\sum_{j=0}^{4}(\gamma_j - \gamma'_{j-k} - m_j)\zeta^{2j} = 0,$$

where the sum in the subscripts is modulo 5. Thus, the vector with jth components $\gamma_j - \gamma'_{j-k} - m_j$ is orthogonal to the vectors $\vec{e}_0, \vec{e}_2, \vec{e}_{-2}$ of the basis (5.1), i.e. it is linear combination of $\vec{e}_1$ and $\vec{e}_{-1}$:

$$\gamma_j - \gamma'_{j-k} - m_j = \mathrm{Re}(z_0\zeta^{-j}) \tag{5.41}$$

for some $z_0 \in \mathbb{C}$. From (5.29), which gives a necessary and sufficient condition for $\vec{n}$ to belong to Λ_γ, we see that $\vec{n} \in \Lambda_\gamma$ if and only if the tuple with jth component $n_{j-k} - m_{j-k}$ belongs to $\Lambda_{\gamma'}$; the first tuple gives the point $\sum_{j=0}^{4} n_j\zeta^j$, and the second the point $\zeta^k\sum_{j=0}^{4}(n_j - m_j)\zeta^j$, proving that (5.40) is satisfied.

Conversely, assume that (5.40) holds. If we define $\gamma''_j := \gamma'_{j-k} + m_j$, the tiling associated with the pentagrid with parameter $\vec{\gamma}''$ coincides with $\mathcal{T}$. It follows from (i) that $\omega_{\gamma''} = \omega_\gamma$, and then $\omega_\gamma - \zeta^{2k}\omega_{\gamma'} = \sum_{j=0}^{4} m_j\zeta^{2j} \in \mathcal{I}$. ∎

5.5 Singular Pentagrids

A main result in [dB81] is that every Penrose tiling can be obtained from the pentagrid construction, provided one takes into account also some singular pentagrids. We will not discuss this aspect here, since it goes beyond the scope of this book. However, we will give an algebraic characterization of singular pentagrids (cf. Theorem 11.1 of [dB81]).

We adopt the same notations of Sect. 5.4. In particular, if $\vec{\gamma} \in \mathbb{R}^5$, we denote by ω_γ the corresponding complex parameter (5.37).

Proposition 5.19 *A pentagrid with parameter $\vec{\gamma}$ satisfying* (5.9) *is singular if and only if*

$$\omega_\gamma = \mathrm{i}\lambda\zeta^k + p \tag{5.42}$$

for some $\lambda \in \mathbb{R}$, $k \in \{0, \dots, 4\}$ and p belonging to the ideal $\mathcal{I}$ in (5.38).

Proof Assume that $\vec{\gamma} \in \mathbb{R}^5$ satisfies (5.9), and let P_γ be the associated pentagrid. To begin with, we want to study what happens to the parameter (5.37), and to the pentagrid, if we transform $\vec{\gamma}$ by suitable translations or by a cyclic permutation. In the following table, $z_0 \in \mathbb{C}$ and $\vec{n} = (n_0, \dots, n_4) \in \mathbb{Z}^5$ satisfies $n_0 + \dots + n_4 = 0$.

Transformation	$\Rightarrow$	Parameter	Pentagrid
(i) $\gamma'_j = \gamma_j + \mathrm{Re}(z_0\zeta^{-j})$		$\omega_{\gamma'} = \omega_\gamma$	$P_{\gamma'} = P_\gamma - z_0$
(ii) $\vec{\gamma}' = (\gamma_1, \dots, \gamma_4, \gamma_0)$		$\omega_{\gamma'} = \zeta^{-2}\omega_\gamma$	$P_{\gamma'} = \zeta^{-1}P_\gamma$
(iii) $\vec{\gamma}' = \vec{\gamma} + \vec{n}$		$\omega_{\gamma'} = \omega_\gamma + \sum_{j=0}^4 n_j\zeta^{2j}$	$P_{\gamma'} = P_\gamma$

Invariance of (5.37) under the first transformation follows from (5.26), which tells us that, if $\gamma'_j = \gamma_j + \mathrm{Re}(z_0\zeta^{-j})$, then

$$\omega_{\gamma'} - \omega_\gamma = \sum\nolimits_{j=0}^4 \mathrm{Re}(z_0\zeta^{-j})\zeta^{2j} = 0.$$

The rest of the transformation rules in the table are straightforward. In particular, for the last column one simply uses the equations of the pentagrid. Note that in all three cases, P_γ is singular if and only if $P_{\gamma'}$ is singular.

Lines in a pentagrid have directions $(\mathrm{i}\zeta^j)_{j=0}^4$. A pentagrid with parameter $\vec{\gamma}$ is singular if and only if (at least) three lines meet at a point z_0, that means that one line is a symmetry axis for the pair formed by the other two. Using the transformation (i) and (ii), we can reduce the study to the case when the intersection point is 0 and the line of symmetry is the imaginary axis. The new parameter $\vec{\gamma}'$ will have $\gamma'_0 = 0$ and either $\gamma'_1, \gamma'_4 \in \mathbb{Z}$, if the first and fourth grid meet at 0, or $\gamma'_2, \gamma'_3 \in \mathbb{Z}$, if the second and third grid meet at 0. With a transformation (iii) we further reduce to one of the two cases $\gamma'_1 = \gamma'_4 = 0$ and $\gamma'_2 = -\gamma'_3$, or $\gamma'_2 = \gamma'_3 = 0$ and $\gamma'_1 = -\gamma'_4$. Both imply that $\omega_{\gamma'}$ is purely imaginary, and since $\omega_\gamma = \omega_{\gamma'}\zeta^k + p$ for some k and with

$$p = -\sum_{j=0}^{4} n_j\zeta^{2j} = -(n_0\zeta^0 + n_3\zeta + n_1\zeta^2 + n_4\zeta^3 + n_2\zeta^4) \in \mathcal{I},$$

we get (5.42).

Conversely, assume that (5.42) holds. With the transformations (i)–(iii) we get a new pentagrid with complex parameter $\omega_{\gamma'} = \mathrm{i}\lambda$, $\lambda \in \mathbb{R}$. From (5.2) we deduce that

$$\gamma'_j = \frac{2}{5}\mathrm{Re}(\mathrm{i}\lambda\zeta^{-2j}) + \mathrm{Re}(z_0\zeta^{-j})$$

where z_0 can be changed arbitrarily with a transformation (i). If we choose

$$z_0 = -\frac{2}{5}\mathrm{i}\lambda\frac{\sin(4\pi/5)}{\sin(2\pi/5)},$$

we get $\gamma_1' = \gamma_4' = 0$ and $\gamma_2' = -\gamma_3'$, which implies that the grids 0, 1 and 4 meet at the origin. ■

Chapter 6
The Noncommutative Space of Penrose Tilings

In Chap. 3 we saw that Penrose tilings are parameterized by index sequences, that are binary sequences satisfying the grammar rule (3.19). We saw that congruence classes of Penrose tilings (with either Robinson triangles, kites and darts, or golden rhombi) are in bijection with classes of index sequences modulo tail equivalence. In this chapter we want to have a closer look at this space parameterizing Penrose tilings, and understand a bit better its structure. Following A. Connes, we will argue that the natural language to study this space is that of Noncommutative Geometry [Con94]. The purpose of this chapter is to sketch the basic ideas.

Very close to the above-mentioned space is the quotient of the set $2^{\mathbb{N}}$ of all binary sequences by tail equivalence. Let us denote by $\mathbb{S}$ the set of index sequences, by R_{tail} the relation of tail equivalence on $2^{\mathbb{N}}$, and by $\widetilde{R}_{\text{tail}}$ its restriction to $\mathbb{S}$.

There is a bijection $2^{\mathbb{N}} \to \mathbb{S}$ consisting in replacing every "1" in a binary sequence by a string "10". This bijection, however, does not transform R_{tail} into $\widetilde{R}_{\text{tail}}$.

It is natural to wonder whether the two quotient sets

$$\mathbb{S}/\widetilde{R}_{\text{tail}} \qquad \text{and} \qquad 2^{\mathbb{N}}/R_{\text{tail}} \tag{6.1}$$

are "the same", in some sense. And if they are different, how do we "measure" their difference?

The set $2^{\mathbb{N}}$, thought as an infinite product of discrete 2-point spaces with product topology, is homeomorphic to the Cantor set. One can then expect that the quotients in (6.1) are rather complicated objects.

Note that the bijection $2^{\mathbb{N}} \to \mathbb{S}$ mentioned above is, in fact, a homeomorphism if we equip $\mathbb{S}$ with the subspace topology from $2^{\mathbb{N}}$, so that the space of index sequences is homeomorphic to the Cantor set as well.

A naive approach is to consider the sets in (6.1) as topological spaces with quotient topology. However, in both cases, equivalence classes are dense and the quotient topology is the indiscrete one: the only open sets are $\varnothing$ and the whole space. And

F. D'Andrea, *A Guide to Penrose Tilings*,
https://doi.org/10.1007/978-3-031-28428-1_6

since the spaces in (6.1) have the same cardinality (the cardinality of $\mathbb{R}$), they are of course homeomorphic.

In the spirit of moduli spaces, a better approach is to think of an equivalence relation R on a topological space X as a (topological) groupoid.

The slogan now is the following:

1. "Bad" quotient spaces (e.g. non Hausdorff) may be "nice" groupoids (e.g. étale). The caveat is that in order to get a nice groupoid, one has to put a topology on R which is not the subspace topology from $X \times X$.
2. "Nice" groupoids can be efficiently studied using C*-algebras.
3. C*-algebras can be distinguished using K-theory (or a more refined invariant, like Elliot's invariant for AF algebras [Ell76]).

Here is where the point of view of Noncommutative Geometry enters into the game. We shall focus on compact Hausdorff spaces, such as the Cantor set above, and if X is such a space, we will denote by $C(X)$ the algebra of complex-valued continuous functions on X. By Gelfan'd theorem, the contravariant functor C — which sends X to $C(X)$ and a continuous map $f : X \to Y$ to its pullback $f^* : C(Y) \to C(X)$— is an equivalence between the category of compact Hausdorff spaces and that of unital *commutative* C*-algebras. This validates the conceptual point of view that a noncommutative C*-algebra should be regarded as dual to a virtual "noncommutative topological space". The archetype of noncommutative space is the phase space of quantum mechanics, where position and momentum are represented by operators satisfying the canonical commutation relations (CCR). This is why, sometimes, instead of "noncommutative space" some authors prefer the term "quantum space". Another important example from quantum physics is the *CAR algebra*, describing fermions and whose generators satisfy the canonical anticommutation relations.

The correspondence between topological spaces and C*-algebras can be extended to more general geometric objects, namely groupoids [Ren06], and in particular to equivalence relations. Given a "nice" equivalence relation (for example, locally compact Hausdorff and étale), we can transform it into a *-algebra with convolution product, and then complete it to a C*-algebra (the *full groupoid C*-algebra*: the reason for this choice will be clear later on). Even if this correspondence is not functorial, it transforms Morita equivalent groupoids (in the sense of Renault) into Morita equivalent C*-algebras (in the sense of Rieffel). In the special case of commutative C*-algebras, i.e. topological spaces, Morita equivalence is the same as isomorphism. We might then argue that we should regard the C*-algebra as the true geometric object, and that two quotient spaces should be regarded as "the same" if their C*-algebras are Morita equivalent. For "nice" quotient spaces X/R (e.g. when the equivalence classes are the orbits of a group acting freely and properly on X, cf. "situation 2" in [Rie82]), the groupoid C*-algebra is Morita equivalent to $C(X/R)$, further validating the above perspective.

The busy reader, now, may want to know what kind of noncommutative algebras are associated to the quotient spaces (6.1). Both are inductive limits of sequences of finite-dimensional C*-algebras, and can be distinguished using their scaled dimension group, which is a complete invariant for AF algebras. The C*-algebra of the space $\mathbb{S}/\widetilde{R}_{\text{tail}}$ of Penrose tilings is computed in [Con94, Sect. 2.3], and is the one called *Fibonacci C*-algebra* in [Dav96]. The groupoid C*-algebra of the quotient $2^{\mathbb{N}}/R_{\text{tail}}$ turns out to be the CAR algebra mentioned before [Ren06, Example 1.10]. Since K_0 of the Fibonacci algebra is a free abelian group with 2 generators, and K_0 of the CAR algebra is the group of dyadic rationals [Dav96], this is enough to conclude that the two C*-algebras are not Morita equivalent.

In the next sections we develop some of the ideas just described, and give a few details of the constructions involved. Ideally, this chapter is aimed at readers with a basic knowledge of C*-algebras and K-theory or, even better, with a basic knowledge of Noncommutative Geometry (for an introduction to this topic, a good starting point is [GBVF13, Lan03]).

6.1 Topology of the Space of Penrose Tilings

6.1.1 The Cantor Set

We start by recalling the familiar construction of the Cantor set $\mathcal{C}$, as can be found in many textbooks of real analysis. By definition, $\mathcal{C} = \bigcap_{n\geq 0} \mathcal{C}_n$ is the subset of $\mathbb{R}$ defined recursively by $\mathcal{C}_0 = [0, 1]$ and, for all $n \geq 1$,

$$\mathcal{C}_n = \frac{1}{3}\big(\mathcal{C}_{n-1} \cup (\mathcal{C}_{n-1} + 2)\big).$$

For every $n \geq 1$, $\mathcal{C}_n$ is a union of disjoint closed intervals obtained from the intervals of $\mathcal{C}_{n-1}$ by removing their "middle-third" open interval, cf. Fig. 6.1.

On the Cantor set, we will consider the subspace topology inherited from the standard topology of $\mathbb{R}$. With this topology, $\mathcal{C}$ is compact (closed in [0, 1])

Note that a real number in [0, 1] belongs to $\mathcal{C}_1$ if and only if its first digit in base 3 is different from 1, it belongs to $\mathcal{C}_2$ if also the second digit is different from 1, etc. Thus, $\mathcal{C}$ is the set of real numbers in [0, 1] that in base 3 have no digit equal to 1.

Next, let $2^{\mathbb{N}} = \prod_{n=0}^{\infty}\{0, 1\}$ be the set of binary sequences, with lexicographic order and Tychonoff topology (the topology on $\{0, 1\}$ being the discrete one). A metric on $2^{\mathbb{N}}$ is given by

Fig. 6.1 The Cantor set

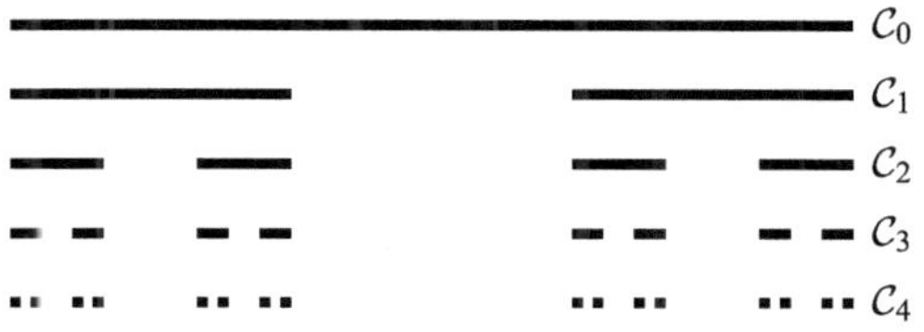

$$d(a,b) := \inf\{2^{-n} : n \in \mathbb{N} \text{ and } a_i = b_i \ \forall\, 0 \le i < n\}, \tag{6.2}$$

for all $a := (a_n)_{n\ge 0}$ and $b := (b_n)_{n\ge 0}$ in $2^{\mathbb{N}}$. By convention, the inf is 1 if $a_0 \ne b_0$.

One easily checks that the open balls are the whole $2^{\mathbb{N}}$ and the cylinder sets

$$\mathbb{B}(x,n) := \{a = (a_k)_{k\ge 0} \in 2^{\mathbb{N}} : a_i = x_i \ \forall\, 0 \le i < n\}, \tag{6.3}$$

where $n \ge 1$ and $x = (x_0, \dots, x_{n-1}) \in \{0,1\}^n$. As a consequence, the metric topology and the Tychonoff topology on $2^{\mathbb{N}}$ coincide. It is instructive to prove that the obvious map from $2^{\mathbb{N}}$ to $\mathcal{C}$ is a homeomorphism.

Lemma 6.1 *The map $f : 2^{\mathbb{N}} \to \mathcal{C}$ given by*

$$f(a) := \sum_{n\ge 0} \frac{2a_n}{3^{n+1}}$$

is a homeomorphism.

Proof Surjectivity of f is obvious. Assume that $a > b$. This means that, for some $k_0 \ge 0$, one has $a_k = b_k$ for all $k < k_0$, and $a_{k_0} = 1 > b_{k_0} = 0$. Clearly $d(a,b) = 2^{-k_0}$. Since

$$f(a) - f(b) = \frac{2}{3^{k_0+1}}\Big(1 + \sum_{n>k_0} \frac{a_n - b_n}{3^{n-k_0}}\Big) \ge \frac{2}{3^{k_0+1}}\Big(1 - \sum_{n\ge 1} \frac{1}{3^n}\Big) = \frac{1}{3^{k_0+1}} > 0,$$

the map f is strictly increasing, and then injective. From the absolute convergence and using $|a_n - b_n| \le 1$, we get

$$d(f(a), f(b)) \le \sum_{n\ge 0} \left| \frac{2a_n}{3^{n+1}} - \frac{2b_n}{3^{n+1}} \right| \le \sum_{n\ge k_0} \frac{1}{2^n} = 2^{-k_0} \sum_{n\ge 0} \left(\tfrac{1}{2}\right)^n = 2\, d(a,b),$$

proving that f is Lipschitz continuous, hence continuous. Since the domain of f is compact (by Tychonoff's theorem), f is a homeomorphism. ∎

6.1.2 The Space of Penrose Tilings

As before, we denote by $\mathbb{S}$ be the set of binary sequences $a = (a_k)_{k\in\mathbb{N}}$ satisfying the grammar rule (3.19):

$$\forall\, i \ge 0, \quad a_i = 1 \implies a_{i+1} = 0.$$

We equip this set with the subspace topology inherited from $2^{\mathbb{N}}$.

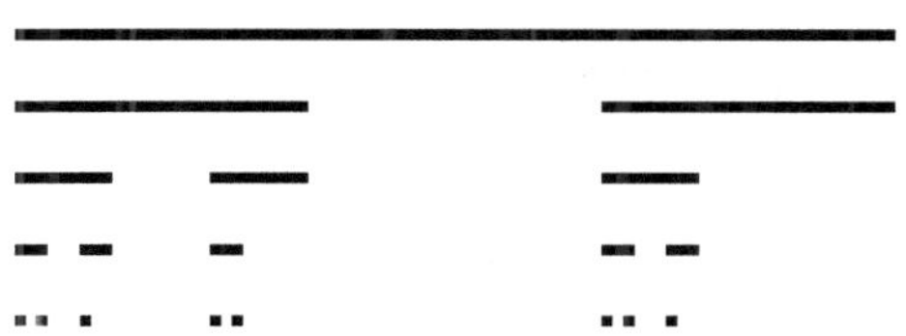

Fig. 6.2 The space $\mathbb{S}$

Its image under the map in Lemma 6.1 is an infinite intersection of closed subsets of [0, 1]. The first sets of this intersection are in Fig. 6.2.

Proposition 6.2 *The map* $\Phi : 2^{\mathbb{N}} \to \mathbb{S}$ *replacing every* 1 *in a binary sequence by a pair* 10 *is a homeomorphism.*

Proof The map Φ is bijective by definition of $\mathbb{S}$: the inverse map applied to a sequence in $\mathbb{S}$ removes the first 0 after each 1. Since the cylinder sets (6.3) form a basis for the topology of $2^{\mathbb{N}}$, it is enough to observe that Φ maps cylinder sets to cylinder sets. More precisely $\Phi\big(U(x,n)\big) = U(x', n')$, where $x' \in \{0,1\}^{n'+1}$ is obtained from x by inserting a 0 after each 1. ∎

Let $R_{\text{tail}} \subset 2^{\mathbb{N}} \times 2^{\mathbb{N}}$ be the relation that we called "tail equivalence" in Sect. 3.4.2. Explicitly, this is an increasing union

$$R_{\text{tail}} := \bigcup_{n \in \mathbb{N}} \big\{(a,b) \in 2^{\mathbb{N}} \times 2^{\mathbb{N}} : a_i = b_i \ \forall\, i \geq n\big\}.$$

Let $\widetilde{R}_{\text{tail}} := R_{\text{tail}} \cap (\mathbb{S} \times \mathbb{S})$ its restriction to $\mathbb{S}$.

It is easy to see that the map in Proposition 6.2 does not preserve tail equivalence. Given the two eventually-periodic sequences (with all entries but the first equal to 1):

$$a = (0, 1, 1, \ldots), \qquad b = (1, 1, 1, \ldots),$$

one computes

$$\Phi(a) = (0, 1, 0, 1, 0, 1, \ldots), \qquad \Phi(b) = (1, 0, 1, 0, 1, 0, \ldots).$$

Obviously, $(a, b) \in R_{\text{tail}}$, but $(\Phi(a), \Phi(b)) \notin \widetilde{R}_{\text{tail}}$, since $\Phi(a)_i = 0$ and $\Phi(b)_i = 1$ for all $i \in 2\mathbb{N}$. Similarly, consider the sequences

$$a = (0, 0, 0, 1, 0, 1, 0, \ldots), \qquad b = (1, 0, 1, 0, 1, 0, \ldots),$$

both with 10 repeated endlessly. Since $a_i = 0$ and $b_i = 1$ for all $i \in 2\mathbb{N}$, $(a, b) \notin R_{\text{tail}}$. On the other hand the sequences

$$\Phi(a) = (0, 0, 0, 1, 0, 0, 1, 0, 0, \ldots), \qquad \Phi(b) = (1, 0, 0, 1, 0, 0, 1, 0, 0, \ldots),$$

are tail equivalent.

The quotient spaces (6.1) share the property of being both indiscrete (with the quotient topology). This follows from the next property of the equivalence relations.

Definition 6.3 An equivalence relation on a topological space X is called *minimal* if its equivalence classes are dense in X.

Proposition 6.4 *Both* R_{tail} *and* $\widetilde{R}_{\mathrm{tail}}$ *are minimal.*

Proof We denote by $[a]$ the class of $a \in 2^{\mathbb{N}}$ under tail equivalence and by $\mathbb{B}(x, n)$ the basic open sets (6.3). For all $n \geq 1$ and $x \in \{0, 1\}^n$:

$$(x_0, x_1, \ldots, x_{n-1}, 0, a_{n+1}, a_{n+2}, \ldots) \in [a] \cap \mathbb{B}(x, n),$$

hence $[a]$ has non-empty intersection with every basic open set, and R_{tail} is minimal. The same works for $\widetilde{R}_{\mathrm{tail}}$: the 0 in place of a_n ensures that (3.19) is satisfied. ■

Now, if a surjective continuous map $f : X \to Y$ has dense fibers, then the topology on Y must be the indiscrete one. Indeed, let U be a proper open subset of Y and let $x \in X$ such that $f(x) \notin U$. The open set $f^{-1}(U)$ has empty intersection with $f^{-1}(x)$, but since the latter set is dense in X, it must be $f^{-1}(U) = \varnothing$, and then $U = \varnothing$.

In particular, if R is a minimal equivalence relation on a topological space X, the quotient topology on X/R is the indiscrete one.

A pair (X, R), where X is homeomorphic to the Cantor set and R is a minimal equivalence relation on X, is called a *Cantor minimal system*. We will see more examples of Cantor minimal systems in Sect. 6.4. Before that, we need to take a detour on inductive limits of finite-dimensional C*-algebras and their K-theory.

6.2 The Canonical Anticommutation Relations

The canonical anticommutation relations arise when one studies a physical system of identical particles obeying the Fermi-Dirac statistics. If the state of a single particle is described by a ray in a complex separable Hilbert space $\mathcal{H}$, a state of n particles will be described by a ray in the antisymmetric Hilbert space tensor product $\bigwedge^n \mathcal{H}$. The wedge product $c(f) := f \wedge \_$ by a vector $f \in \mathcal{H}$ (the tensor product, composed with the orthogonal projection $\bigotimes^n \mathcal{H} \to \bigwedge^n \mathcal{H}$) defines a bounded operator $c(f)$ on the fermionic Fock space $\mathcal{F}(\mathcal{H}) := \bigoplus_{n \geq 0} \bigwedge^n \mathcal{H}$. This is the so-called *creation operator*, which "creates" an extra fermion in the state f. Its adjoint $c(f)^*$ "destroys" the fermion and is called *annihilation operator*. The map c from $\mathcal{H}$ to bounded operators on $\mathcal{F}(\mathcal{H})$ is linear and transforms the wedge product into the composition of operators, which is then skew-symmetric:

$$c(f)c(g) + c(g)c(f) = 0, \qquad \forall\, f, g \in \mathcal{H}. \tag{6.4a}$$

With an explicit computation one also proves that

$$c(f)^*c(g) + c(g)c(f)^* = \langle f, g \rangle\, I, \qquad \forall\, f, g \in \mathcal{H}. \tag{6.4b}$$

On the right hand side of (6.4b), I is the identity operator on $\mathcal{F}(\mathcal{H})$ and the convention on the inner product is that it is antilinear in the left entry. The equations displayed in (6.4) are called Canonical Anticommutation Relations (CAR).

From now on, we assume that $\mathcal{H} = \ell^2(\mathbb{N})$ is (separable and) infinite-dimensional. The *CAR algebra* is the norm-closure of the unital *-algebra generated by the operators $\{c(f), c(f)^* : f \in \mathcal{H}\}$. We will see that its structure depends only on (6.4), and not the explicit form of the map c.

First, if $f \in \mathcal{H}$ is a unit vector, using (6.4) with $g = f$ yields $c(f)^2 = 0$ and $c(f)^*c(f) + c(f)c(f)^* = I$. Multiplying the latter equation by $c(f)$ yields

$$c(f)c(f)^*c(f) = c(f). \tag{6.5}$$

That is, $c(f)$ is a partial isometry. We associate to f three operators:

$$P(f) := c(f)c(f)^*, \qquad P(f)^\perp := c(f)^*c(f), \qquad \gamma(f) := P(f) - P(f)^\perp.$$

The first two are projections: $P(f)$ is the orthogonal projection onto the range of $c(f)$ and $P(f)^\perp = I - P(f)$. The last operator is a $\mathbb{Z}_2$-grading, i.e. $\gamma(f) = \gamma(f)^*$ and $\gamma(f)^2 = I$, and anticommutes with $c(f)$ and $c(f)^*$, as one can check using (6.5).

Lemma 6.5 *If f, g are orthogonal unit vectors, then*

(i) $c(g)$ and $c(g)^*$ anticommute with both $c(f)$ and $c(f)^*$;
(ii) $P(g)$, $P(g)^\perp$ and $\gamma(g)$ commute with the algebra generated by $c(f)$ and $c(f)^*$.

Proof (i) follows from (6.4) and $\langle f, g \rangle = 0$. Since $P(g)$ is a product of two operators anticommuting with $c(f)$ and $c(f)^*$, we get (ii). ■

Next, fix an orthonormal basis $\{f_n : n \in \mathbb{N}\}$ of $\mathcal{H}$ (for example, the canonical one) and for $n \geq 0$ define

$$E_{11}^{(n)} := P(f_n), \qquad E_{12}^{(n)} := \Big(\prod_{i=0}^{n-1} \gamma(f_i)\Big) c(f_n),$$

$$E_{21}^{(n)} := c(f_n)^* \Big(\prod_{i=0}^{n-1} \gamma(f_i)\Big), \qquad E_{22}^{(n)} := P(f_n)^\perp.$$

Here by convention an empty product is equal to I. Observe that, since the $\gamma(f_i)$'s are mutually commuting, their order in the above product doesn't matter.

Denote by $M_k(\mathbb{C})$ the (C*-)algebra of $k \times k$ complex matrices, and by $\mathcal{B}$ the C*-algebra of bounded operators on the Fock space $\mathcal{F}(\mathcal{H})$.

Lemma 6.6 *(i) For every $n \geq 0$, the map*

$$\Phi^{(n)} : M_2(\mathbb{C}) \to \mathcal{B}, \qquad \begin{pmatrix} a_{11} & a_{12} \\ a_{21} & a_{22} \end{pmatrix} \mapsto \sum_{i,j=1}^{2} a_{ij}\, E_{ij}^{(n)},$$

*is an injective unital *-homomorphism. (ii) For every $m \neq n$, the subalgebras $\Phi^{(m)}(M_2(\mathbb{C}))$ and $\Phi^{(n)}(M_2(\mathbb{C}))$ of $\mathcal{B}$ are mutually commuting.*

Proof (i) For all $n \geq 0$, the $E_{ij}^{(n)}$'s satisfy the relations of elementary 2×2 matrices, i.e. $E_{11}^{(n)}$ and $E_{22}^{(n)}$ are complementary orthogonal projections, $(E_{12}^{(n)})^* = E_{21}^{(n)}$ and

$$E_{ij}^{(n)} E_{kl}^{(n)} = \delta_{jk} E_{il}^{(n)}. \tag{6.6}$$

The latter can be easily deduced from from Lemma 6.5. For example,

$$E_{12}^{(n)} E_{21}^{(n)} = \Big(\prod_{i=0}^{n-1} \gamma(f_i) \Big) P(f_n) \Big(\prod_{i=0}^{n-1} \gamma(f_i) \Big) = P(f_n) \Big(\prod_{i=0}^{n-1} \gamma(f_i) \Big)^2 = P(f_n) = E_{11}^{(n)},$$

where we used Lemma 6.5(ii) and $\gamma(f_i)^2 = I$.

Injectivity follows easily from the above relations. If $Z := \sum_{i,j=1}^{2} a_{ij}\, E_{ij}^{(n)}$ is zero, then $\sum_{k=1}^{2} E_{ki}^{(n)} Z E_{jk}^{(n)} = a_{ij}$ is also 0, for all $i, j \in \{1, 2\}$.

(ii) From Lemma 6.5(ii), $E_{ii}^{(m)}$ and $E_{jk}^{(n)}$ commute for all i, j, k. It remains to show that $E_{12}^{(m)}$ commutes with $E_{jk}^{(n)}$ for $j \neq k$ (by adjunction, $(E_{12}^{(m)})^*$ will commute with $E_{jk}^{(n)}$ as well), and it is enough to do it for $m < n$. Since $c(f_m)$ and $\gamma(f_i)$ commute for $i \neq m$, and anticommute for $i = m$, we have

$$\begin{aligned} E_{12}^{(m)} E_{12}^{(n)} &= \Big(\prod_{i=0}^{m-1} \gamma(f_i) \Big) c(f_m) \Big(\overbrace{\prod_{i=0}^{n-1} \gamma(f_i)}^{(i \text{ may be equal to } m)} \Big) c(f_n) \\ &= -\Big(\prod_{i=0}^{m-1} \gamma(f_i) \Big) \Big(\prod_{i=0}^{n-1} \gamma(f_i) \Big) c(f_m) c(f_n) \\ &= \Big(\prod_{i=0}^{n-1} \gamma(f_i) \Big) \underbrace{\Big(\prod_{i=0}^{m-1} \gamma(f_i) \Big)}_{(\text{here } i \neq n)} c(f_n) c(f_m) \\ &= \Big(\prod_{i=0}^{n-1} \gamma(f_i) \Big) c(f_n) \Big(\prod_{i=0}^{m-1} \gamma(f_i) \Big) c(f_m) = E_{12}^{(n)} E_{12}^{(m)}. \end{aligned}$$

The same computation can be repeated for $E_{21}^{(n)}$, with $c(f_n)$ replaced by $c(f_n)^*$. ■

Let $A_0 := \mathbb{C}I$ be the algebra of scalar multiples of the identity and, for $n \geq 1$, call A_n the algebra generated by the set $\big\{ c(f_k), c(f_k)^* : 0 \leq k \leq n-1 \big\}$. We have an ascending sequence of C*-subalgebras

$$A_0 \subseteq A_1 \subseteq A_2 \subseteq \ldots \subseteq A_n \subseteq \ldots$$

whose union is dense in the CAR C*-algebra (since it contains all of its generators).

For $m \geq 1$, $a \in M_m(\mathbb{C})$ and $b = (b_{ij}) \in M_2(\mathbb{C})$ define:

$$a \otimes \begin{pmatrix} b_{11} & b_{12} \\ b_{21} & b_{22} \end{pmatrix} := \left(\begin{array}{c|c} b_{11}a & b_{12}a \\ \hline b_{21}a & b_{22}a \end{array}\right),$$

where each block on the right is the product of an $m \times m$ matrix and a scalar. (In other words, we identify the tensor product $M_m(\mathbb{C}) \otimes M_2(\mathbb{C})$ with $M_{2m}(\mathbb{C})$.)

Proposition 6.7 *For every $n \geq 1$, the map*

$$\Phi_n : M_{2^n}(\mathbb{C}) = \overbrace{M_2(\mathbb{C}) \otimes \ldots \otimes M_2(\mathbb{C})}^{n \text{ times}} \to \mathcal{B},$$

$$a^{(0)} \otimes a^{(1)} \otimes \ldots \otimes a^{(n-1)} \mapsto \Phi^{(0)}(a^{(0)})\Phi^{(1)}(a^{(1)}) \ldots \Phi^{(n-1)}(a^{(n-1)}),$$

*is an injective unital *-homomorphism whose image is A_n.*

Proof It follows from Lemma 6.6 that Φ_n is a unital *-homomorphism. Clearly $\mathrm{Im}(\Phi_n) \subseteq A_n$. For all $0 \leq k \leq n-1$, $\gamma(f_k) = E_{11}^{(k)} - E_{22}^{(k)}$ belongs to $\mathrm{Im}(\Phi_n)$. Thus,

$$c(f_k) = \Big(\prod_{i=0}^{k-1} \gamma(f_i)\Big) E_{12}^{(k)} \quad \text{and} \quad c(f_k)^* = E_{21}^{(k)} \Big(\prod_{i=0}^{k-1} \gamma(f_i)\Big)$$

belong to $\mathrm{Im}(\Phi_n)$ as well. Since $\mathrm{Im}(\Phi_n)$ contains all the generators of A_n, the map $\Phi_n : M_2(\mathbb{C}) \otimes \ldots \otimes M_2(\mathbb{C}) \to A_n$ is surjective.

Injectivity can be proved by induction on n. For $n = 0$ it follows from Lemma 6.6. Assume, then, that Φ_{n-1} is injective for some $n \geq 1$. Denote by E_{ij} the elementary 2×2 matrices. An element in the domain of Φ_n is a sum $B := \sum_{ij} B_{ij} \otimes E_{ij}$, with B_{ij} in the domain of Φ_{n-1}, and by construction $\Phi_n(B) = \sum_{ij} \Phi_{n-1}(B_{ij})E_{ij}^{(n-1)}$. From (6.6) and the fact that $E_{ij}^{(n-1)}$ commutes with the image of Φ_{n-1}, we get

$$\sum_{k=1,2} E_{ki}^{(n-1)} \Phi_n(B) E_{jk}^{(n-1)} = \Phi_{n-1}(B_{ij}).$$

Thus, $\Phi_n(B) = 0$ if and only if $\Phi_{n-1}(B_{ij}) = 0$ for all i, j, and by inductive hypothesis this means that $B_{ij} = 0$ for all i, j, and so $B = 0$. ∎

Note that the subalgebra A_{n-1} of A_n is the image under Φ_n of tensors with the last leg $a^{(n-1)}$ equal to the 2×2 identity matrix I_2. If we use Φ_n to identify A_n with $M_{2^n}(\mathbb{C})$, we see that the inclusion of A_{n-1} into A_n becomes the map $a \mapsto a \otimes I_2$.

Remark 6.8 The CAR algebra has a dense subalgebra isomorphic to the increasing union $\bigcup_{n\geq 0} M_{2^n}(\mathbb{C})$, where the inclusion $M_{2^{n-1}}(\mathbb{C}) \to M_{2^n}(\mathbb{C})$ is block diagonal:

$$a \mapsto \begin{pmatrix} a & 0 \\ 0 & a \end{pmatrix}.$$

6.3 Approximately Finite-Dimensional C*-algebra

The CAR algebra is the prototype of AF C*-algebra. A C*-algebra is called *approximately finite-dimensional* (AF) if it is the closure of an increasing union of finite-dimensional C*-subalgebras. In this section we will sketch some basic properties of AF algebras, and refer to the textbooks for a detailed study (e.g. [Dav96]).

If A is a unital *-algebra and $n \geq 1$, we denote by $M_n(A)$ the unital *-algebra of $n \times n$ matrices with entries in A, with the obvious unit, involution and multiplication. For all $k, n \geq 1$, we will identify $M_k\big(M_n(A)\big)$ with $M_{kn}(A)$ in the obvious way, "removing parenthesis". For example, the matrix

$$\begin{pmatrix} \begin{pmatrix} a_{11} & a_{12} \\ a_{21} & a_{22} \end{pmatrix} & \begin{pmatrix} b_{11} & b_{12} \\ b_{21} & b_{22} \end{pmatrix} \\ \begin{pmatrix} c_{11} & c_{12} \\ c_{21} & c_{22} \end{pmatrix} & \begin{pmatrix} d_{11} & d_{12} \\ d_{21} & d_{22} \end{pmatrix} \end{pmatrix} \in M_2\big(M_2(A)\big)$$

will be identified with

$$\begin{pmatrix} a_{11} & a_{12} & b_{11} & b_{12} \\ a_{21} & a_{22} & b_{21} & b_{22} \\ c_{11} & c_{12} & d_{11} & d_{12} \\ c_{21} & c_{22} & d_{21} & d_{22} \end{pmatrix} \in M_4(A).$$

In the present chapter we will tacitly assume, unless stated otherwise, that

all algebras and morphisms are unital.

6.3.1 Morphisms of Finite-Dimensional C-Algebras*

By the structure theorem for finite-dimensional C*-algebras, every such algebra is isomorphic to a direct sum of matrix algebras.

A morphism

$$A_1 \cong M_{n_1}(\mathbb{C}) \oplus \ldots \oplus M_{n_k}(\mathbb{C}) \longrightarrow A_2 \cong M_{m_1}(\mathbb{C}) \oplus \ldots \oplus M_{m_\ell}(\mathbb{C}) \tag{6.7}$$

of two such algebras is determined (up to a unitary equivalence in A_2) by an $\ell \times k$ matrix (a_{ij}) of non-negative integers, satisfying

$$\sum_{j=1}^{k} a_{ij} n_j = m_i \qquad \forall\, i = 1, \ldots, \ell. \tag{6.8}$$

This is called the matrix of *partial multiplicities* of the morphism (6.7), and is obtained as follows.

First, every irreducible representation of $M_{n_j}(\mathbb{C})$ is unitary equivalent to the identity representation id_{n_j}. Every representation of $M_{n_1}(\mathbb{C}) \oplus \ldots \oplus M_{n_k}(\mathbb{C})$ is then unitary equivalent to $\mathrm{id}_{n_1}^{a_1} \oplus \ldots \oplus \mathrm{id}_{n_k}^{a_k}$ for suitable multiplicities $a_1, \ldots, a_k \in \mathbb{N}$ (here $\mathrm{id}_{n_j}^{0}$ is the zero map). Composing the map (6.7) with the projection onto the ith summand, we get a non-degenerate representation on $\mathbb{C}^{m_i}$ of the algebra on the left in (6.7). This is of the form $\mathrm{id}_{n_1}^{a_{i1}} \oplus \ldots \oplus \mathrm{id}_{n_k}^{a_{ik}}$ for suitable $a_{ij} \in \mathbb{N}$. A comparison of dimensions yields the condition (6.8).

The entry (i, j) in the matrix of partial multiplicities gives the multiplicity of the embedding of the summand $M_{n_j}(\mathbb{C})$ of A_1 into the summand $M_{m_i}(\mathbb{C})$ of A_2.

Example 6.9 (*CAR algebra*) For every $n \geq 1$, the morphism $M_{2^{n-1}}(\mathbb{C}) \to M_{2^n}(\mathbb{C})$ in Remark 6.8 has matrix of partial multiplicities

$$\begin{pmatrix} 2 \end{pmatrix} \tag{6.9}$$

(The unique summand in the domain is embedded with multiplicity 2 in the codomain.)

Example 6.10 Let $(F_k)_{k\geq 0}$ be the Fibonacci numbers. For every $k \geq 1$ there is a morphism

$$M_{F_k}(\mathbb{C}) \oplus M_{F_{k+1}}(\mathbb{C}) \longrightarrow M_{F_{k+1}}(\mathbb{C}) \oplus M_{F_{k+2}}(\mathbb{C})$$

whose matrix of partial multiplicities is

$$\begin{pmatrix} 0 & 1 \\ 1 & 1 \end{pmatrix} \tag{6.10}$$

(The condition (6.8) in this case is the recursive equation of Fibonacci numbers.)

Remark 6.11 To represent pictorially the morphism (6.7) we write $(n_1, \ldots, n_k)$ and $(m_1, \ldots, m_\ell)$ in two columns and, for each i, j, draw a_{ij} arrows from n_j to m_i. In the two examples above we get, respectively

$$2^{n-1} \rightrightarrows 2^n$$

and

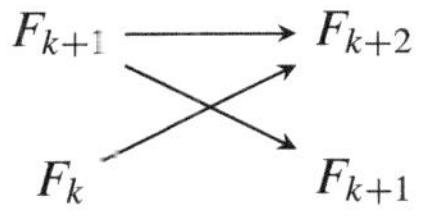

6.3.2 Bratteli Diagrams

By definition a C*-algebra is AF if it is of the form $A = \overline{\bigcup_{n\geq 0} A_n}$, where

$$A_0 \subseteq A_1 \subseteq A_2 \subseteq \ldots \subseteq A_n \subseteq \ldots \tag{6.11}$$

is a nested sequence of finite-dimensional C*-algebras, and we assume that A_0 consists of scalar multiples of the identity. From the discussion in the previous section, it follows that every inclusion $A_{n-1} \to A_n$ is given, up to isomorphisms, by a matrix of partial multiplicities. This sequence of matrices of partial multiplicities is the *Bratteli diagram* of (6.11). It can be represented by an infinite graph as explained in Remark 6.11.

Example 6.12 The Bratteli diagram of the CAR algebra is

$$1 \rightrightarrows 2 \rightrightarrows 4 \rightrightarrows 8 \rightrightarrows 16 \rightrightarrows \cdots$$

One should be careful that different Bratteli diagrams may define the same AF C*-algebra. On the other hand, two AF C*-algebras with the same Bratteli diagram are isomorphic [Dav96, Proposition III.2.7]. The reason is that the Bratteli diagram determines (up to isomorphism) the subalgebra $\bigcup_{n\geq 0} A_n$. But on each finite-dimensional *-algebra A_n there is a unique C*-norm, hence on $\bigcup_{n\geq 0} A_n$ there is a unique C*-norm, and the completion (if it exists) is unique.

To define an AF C*-algebra is then enough to specify its Bratteli diagram.

Example 6.13 We call *Fibonacci C*-algebra* the AF C*-algebra with Bratteli diagram:

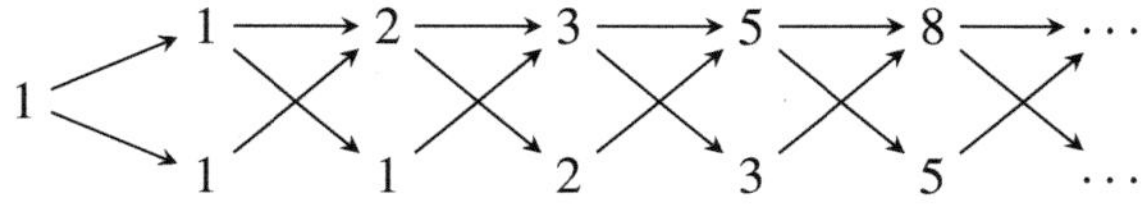

That is, for each $k \geq 1$, the morphism $A_k \to A_{k+1}$ is the one in Example 6.10.

Both the CAR and Fibonacci C*-algebras have the peculiar property that the matrix of partial multiplicities of the inclusion $A_{n-1} \to A_n$ is independent of n (for $n \geq 1$ in the second case).

Observe that in a Bratteli diagram there is no need to write the value of the dimensions of the algebras involved. All the dimensions can be deduced from the diagram recursively, using $\dim(A_0) = 1$. For v a vertex in a Bratteli diagram, let us denote by $\dim(v)$ the dimension of the matrix algebra at that vertex. Then, from (6.8) we deduce the following recursive formula for dimensions:

$$\dim(v) = \sum_{e \in t^{-1}(v)} \dim s(e)$$

where the sum is over edges, and s and t are the source and target maps.

Let us give a formal general definition of Bratteli diagram, since it will be useful in the subsequent sections.

Definition 6.14 A *Bratteli diagram* is a directed graph $\mathfrak{B} = (V, E, s, t)$ whose sets of vertices and edges are countably infinite disjoint unions

$$V = \bigsqcup_{i \geq 0} V_i, \qquad E = \bigsqcup_{i \geq 0} E_i,$$

of finite non-empty sets such that:

(i) $V_0 = \{v_0\}$ is a singleton;
(ii) for all $i \geq 0$, edges in E_i have source in V_i and target in V_{i+1};
(iii) the graph has no sinks and no sources other than v_0, that is:

$$s^{-1}(v) \neq \varnothing \ \forall \ v \in V, \qquad t^{-1}(v) \neq \varnothing \ \forall \ v \in V \setminus V_0.$$

Condition (i) is motivated by the fact that, in our AF algebras, we always have $A_0 = \mathbb{C}I$. Condition (iii) reflects the fact that, in an AF algebra, for every $n \geq 0$, each summand of A_n is embedded in at least one summand of A_{n+1} (no sinks), and each summand of A_{n+1} contains at least one summand of A_n (no sources). That is, they correspond to the fact that each *-homomorphism $A_n \to A_{n+1}$ is injective (no sinks) and unital (no sources).

6.3.3 K-Theory of AF Algebras

A *cone* G^+ in an abelian group G is a sub-monoid satisfying $G^+ \cap (-G^+) = \{0\}$. A cone induces a partial order on G defined by $x \leq y \iff y - x \in G^+$. We shall assume that our ordered groups are *directed*, i.e. that $G = G^+ - G^+$.

A *scale* for a pair (G, G^+) as above is a set Σ of generators of G^+ which is directed (for every $g_1, g_2 \in \Sigma$ there is $g \in \Sigma$ such that $g \geq g_1$ and $g \geq g_2$) and hereditary (if $0 \leq g_0 \leq g_1$ and $g_1 \in \Sigma$, then $g_0 \in \Sigma$). A triple (G, G^+, Σ) as above is called a *scaled directed ordered group* (SDOG).

A *morphism* $(G, G^+, \Sigma) \to (H, H^+, S)$ of SDOGs is a group homomorphism $G \to H$ which maps G^+ to H^+ and Σ to S (this is called a *contractive positive homomorphism* in [Dav96]).

To every AF algebra A, we associate a SDOG

$$\mathbf{K}_0(A) := \big(K_0(A), K_0^+(A), \Sigma(A)\big) \tag{6.12}$$

as follows. Recall that $P \in M_k(A)$ is called a *projection* if

$$P^2 = P = P^*.$$

Two projections $P \in M_k(A)$ and $Q \in M_m(A)$ are (von Neumann) *equivalent* if there is a $k \times m$ matrix U with entries in A such that

$$UU^* = P \quad \text{and} \quad U^*U = Q.$$

There is a weaker notion of "stable equivalence", but for AF algebras von Neumann equivalence and stable equivalence are the same [Dav96, Theorem IV.1.6].

Embedding each matrix algebra $M_k(A)$ in $M_{k+1}(A)$ as upper left corner (observe that this is a non-unital *-homomorphism, in contrast with the maps that we met in the previous sections), we can form the inductive limit $M_\infty(A) := \bigcup_{k\geq 1} M_k(A)$.

$K_0^+(A)$ is defined as the set of equivalence classes of projections in $M_\infty(A)$, and it is an abelian cancellative monoid with operation induced by the direct sum of projections,

$$[P] + [Q] = [P \oplus Q],$$

and with neutral element given by the class [0] of the null projection.

We define $K_0(A)$ as the Grothendieck group of $K_0^+(A)$, and

$$\Sigma(A) := \big\{[P] \in K_0^+(A) : P \in A\big\}.$$

$K_0^+(A)$ is a cone in $K_0(A)$ and $\Sigma(A)$ is a scale, cf. [Dav96, §IV]. In particular, the K_0-group of an AF algebra A is always generated by projections in A. (The K_1-group an AF-algebra is always zero, hence it does not carry any additional information, see e.g. [Bla86, Example 8.1.2].)

Example 6.15 Let $A \cong M_n(\mathbb{C})$. Every projection $P \in M_k(A) \cong M_{kn}(\mathbb{C})$ is unitary equivalent to a diagonal matrix with 0's and 1's on the diagonal (as one can see by diagonalizing it). For every $1 \leq i \leq kn$, if we insert after the top-left $i \times i$ sub-matrix a row and a column full of zeros, we don't change the class of P. With this trick, one shows that P is von Neumann equivalent to a direct sum of r copies of the matrix unit $E_{11} \in A$ (the matrix with 1 in position (1, 1) and zero everywhere else), where r is the rank of P. The map sending $[P]$ to the trace of P defines an isomorphism

$$\mathbf{K}_0(M_n(\mathbb{C})) \cong (\mathbb{Z}, \mathbb{N}, \underline{n} := \{0, 1, \dots, n\}).$$

(When we apply the trace to $\Sigma(A)$ we get the set $\underline{n}$.)

Arbitrary finite-dimensional C*-algebras are not much more complicated.

Example 6.16 Let $A \cong M_{n_1}(\mathbb{C}) \oplus \ldots \oplus M_{n_\ell}(\mathbb{C})$. By shuffling blocks, we can identify $M_k(A)$ with $M_{kn_1}(\mathbb{C}) \oplus \ldots \oplus M_{kn_\ell}(\mathbb{C})$. Every $P \in M_k(A)$ decomposes as a direct sum $P_1 \oplus \ldots \oplus P_\ell$ with $P_i \in M_{kn_i}(\mathbb{C})$ a projection. By the same argument in Example 6.15, the map

$$P_1 \oplus \ldots \oplus P_\ell \to \big(\mathrm{Tr}(P_1), \ldots, \mathrm{Tr}(P_\ell)\big)$$

induces an isomorphism

$$\mathbf{K}_0(A) \cong (\mathbb{Z}^\ell, \mathbb{N}^\ell, \underline{n}_1 \times \ldots \times \underline{n}_\ell)\,. \tag{6.13}$$

Every *-homomorphism $f : A \to B$ of AF C*-algebras induces a morphism $f_* : \mathbf{K}_0(A) \to \mathbf{K}_0(B)$ via the formula

$$f_*([P]) := [f(P)],$$

where $P \in M_k(A)$ is a projection and f is applied to each matrix entry of P. In this way, we get a functor from AF C*-algebras to SDOGs.

Example 6.17 Consider a morphism $f : A_1 \to A_2$ of finite-dimensional C*-algebras like in (6.7), and let $a = (a_{ij})$ be the matrix of partial multiplicities. With an abuse of notations, let us denote by $E_{11} \in M_n(\mathbb{C})$ the matrix unit with 1 in position $(1, 1)$ and zero everywhere else (regardless of the value of n). Given a direct sum of full matrix algebras, let $E_{11}^{(j)}$ be the matrix E_{11} inserted as jth summand. The matrices $\{E_{11}^{(j)}\}_{j=1,\ldots,k}$ generate the K-theory of A_1. The trace map sends $E_{11}^{(j)}$ to the tuple $e_m \in \mathbb{Z}^k$ with 1 in position m and zero every where else. By definition of matrix of partial multiplicities,

$$f_*\big([E_{11}^{(j)}]\big) = \sum_{i=1}^{\ell} a_{ij}[E_{11}^{(i)}].$$

Thus, under the identification (6.13) given by the trace map, f_* becomes the morphism $\mathbb{Z}^k \to \mathbb{Z}^\ell$ given by matrix multiplication by the matrix of partial multiplicities.

Let $\mathbf{G}_i = (G_i, G_i^+, S_i)$ be SDOGs, $i \geq 0$, and suppose that we have a sequence of morphisms

$$\mathbf{G}_0 \xrightarrow{f_0} \mathbf{G}_1 \xrightarrow{f_1} \mathbf{G}_2 \longrightarrow \ldots \longrightarrow \mathbf{G}_i \xrightarrow{f_i} \mathbf{G}_{i+1} \longrightarrow \ldots \tag{6.14}$$

This means that we have a sequence of morphims of abelian groups, which restricts to a sequence of maps between positive cones, which restricts to a sequence of maps between scales. The inductive limit of these three sequences gives a triple

$$(G, G^+, S) := \varinjlim \mathbf{G}_i := \big(\varinjlim G_i, \varinjlim G_i^+, \varinjlim S_i\big).$$

Explicitly, as a set

$$G = \big\{(i, g) : i \in \mathbb{N}, g \in G_i\big\}/\!\sim$$

where $(i_1, g_1) \sim (i_2, g_2)$ if there exists $j \geq i_1, i_2$ such that

$$f_{j-1} \circ \ldots \circ f_{i_1+1} \circ f_{i_1}(g_1) = f_{j-1} \circ \ldots \circ f_{i_2+1} \circ f_{i_2}(g_2).$$

The convention here is that an empty composition gives the identity map. Given two elements $[(i, g)], [(j, h)] \in G$, with $i \geq j$, their sum is defined as

$$[(i, g)] + [(j, h)] := [(g + f_{i-1} \circ \ldots \circ f_j(h), i)].$$

Replacing each G_i with G_i^+ (resp. S_i) in the above construction we get G^+ (resp. S).

A crucial point for our next computations is that, if the maps f_i are inclusions and

$$\mathbf{H} = \left(\bigcup_{i \geq 0} G_i , \bigcup_{i \geq 0} G_i^+ , \bigcup_{i \geq 0} S_i \right)$$

is a SDOG, then $\varinjlim \mathbf{G}_i \cong \mathbf{H}$. Another important point is that, if we have a commutative diagram of SDOGs whose vertical arrows are isomorphisms:

$$\begin{array}{ccccccccccc} \mathbf{G}_0 & \longrightarrow & \mathbf{G}_1 & \longrightarrow & \mathbf{G}_2 & \longrightarrow & \mathbf{G}_3 & \longrightarrow & \mathbf{G}_4 & \longrightarrow & \cdots \\ \downarrow \wr\wr & & \downarrow \wr\wr & & \downarrow \wr\wr & & \downarrow \wr\wr & & \downarrow \wr\wr & & \\ \mathbf{H}_0 & \longrightarrow & \mathbf{H}_1 & \longrightarrow & \mathbf{H}_2 & \longrightarrow & \mathbf{H}_3 & \longrightarrow & \mathbf{H}_4 & \longrightarrow & \cdots \end{array}$$

then the two rows have the same inductive limit.

The next theorem, that we cite without proof (see e.g. [Dav96, Theorem IV.3.3] for the details), is the final ingredient we need to compute the K-theory of AF-algebras.

Theorem 6.18 *If* $A = \overline{\bigcup_{n \geq 0} A_n}$ *is an AF algebra, then*

$$\mathbf{K}_0(A) = \varinjlim \mathbf{K}_0(A_n).$$

An ordered abelian group that is an inductive limit of finitely-generated free abelian groups is called a *dimension group*. The previous theorem and Example 6.16 tell us that, if A is an AF algebra, the pair $(K_0(A), K_0^+(A))$ is a dimension group. Of particular interest are dimension groups defined by *stationary inductive sequences*, i.e. sequences (6.14) where all the $\mathbf{G}_i$'s are equal, and all the maps f_i's are equal. The CAR algebra belongs to this class of examples. The sequence of the Fibonacci algebra is stationary except for the first map.

In what follows, if $r \in \mathbb{R}$, we denote by $\mathbb{Z}[r]$ the abelian group of polynomials with integer coefficients evaluated at r.

Example 6.19 Let A be the CAR C*-algebra. Its Bratteli diagram is in Example 6.12, and the matrix of partial multiplicity is in (6.9). By the discussion in Example 6.17, the sequence defining $\mathbf{K}_0(A)$ is the first row of the following commutative diagram

$$\begin{array}{ccccccccccc} \mathbb{Z} & \xrightarrow{2} & \mathbb{Z} & \xrightarrow{2} & \mathbb{Z} & \xrightarrow{2} & \mathbb{Z} & \xrightarrow{2} & \mathbb{Z} & \longrightarrow & \cdots \\ {\scriptstyle 1}\downarrow & & {\scriptstyle \frac{1}{2}}\downarrow & & {\scriptstyle \frac{1}{2^2}}\downarrow & & {\scriptstyle \frac{1}{2^3}}\downarrow & & {\scriptstyle \frac{1}{2^4}}\downarrow & & \\ \mathbb{Z} & \xrightarrow{1} & \frac{1}{2}\mathbb{Z} & \xrightarrow{1} & \frac{1}{2^2}\mathbb{Z} & \xrightarrow{1} & \frac{1}{2^3}\mathbb{Z} & \xrightarrow{1} & \frac{1}{2^4}\mathbb{Z} & \longrightarrow & \cdots \end{array}$$

By an arrow "$\xrightarrow{\lambda}$" we mean the map given by multiplication by λ. Since the vertical arrows are isomorphisms, $\mathbf{K}_0(A)$ is isomorphic to the direct limit of the second row, which is $\mathbb{Z}[\frac{1}{2}]$. The positive cone is the subset of polynomials with non-negative coefficients evaluated at $1/2$. Finally, in the first row the scale at the site n is given by the subset of integers between 0 and 2^n. Multiplied by $1/2^n$ we obtain the set of $x \in \frac{1}{2^n}\mathbb{Z}$ such that $0 \le x \le 1$. Therefore, we conclude that

$$\mathbf{K}_0(A) \cong \Big(\mathbb{Z}[\tfrac{1}{2}]\,,\ \mathbb{Z}[\tfrac{1}{2}] \cap [0, \infty[\,,\ \mathbb{Z}[\tfrac{1}{2}] \cap [0, 1]\Big). \tag{6.15}$$

The computation for the Fibonacci C*-algebra is a bit more involved, and we present it in the form of a proposition. Recall that φ denotes the golden ratio.

Proposition 6.20 *Let A be the Fibonacci C*-algebra. Then*

$$\mathbf{K}_0(A) \cong \Big(\mathbb{Z}[\varphi]\,,\ \mathbb{Z}[\varphi] \cap [0, \infty[\,,\ \mathbb{Z}[\varphi] \cap [0, 1]\Big). \tag{6.16}$$

Proof The Bratteli diagram is in Example 6.13, and the matrix of partial multiplicities is in (6.10). The sequence defining $\mathbf{K}_0(A)$ is the first row of the following commutative diagram

$$\begin{array}{ccccccccccc}
\mathbb{Z} & \xrightarrow{\binom{1}{1}} & \mathbb{Z}^2 & \xrightarrow{\left(\begin{smallmatrix}0&1\\1&1\end{smallmatrix}\right)} & \mathbb{Z}^2 & \xrightarrow{\left(\begin{smallmatrix}0&1\\1&1\end{smallmatrix}\right)} & \mathbb{Z}^2 & \xrightarrow{\left(\begin{smallmatrix}0&1\\1&1\end{smallmatrix}\right)} & \mathbb{Z}^2 & \longrightarrow & \cdots \\
\Big\downarrow{\scriptstyle \varphi} & & \Big\downarrow{\scriptstyle (\varphi^{-1}\ 1)} & & \Big\downarrow{\scriptstyle (\varphi^{-2}\ \varphi^{-1})} & & \Big\downarrow{\scriptstyle (\varphi^{-3}\ \varphi^{-2})} & & \Big\downarrow{\scriptstyle (\varphi^{-4}\ \varphi^{-3})} & & \\
\varphi\mathbb{Z} & \xrightarrow{\text{id}} & \mathbb{Z}[\varphi] & \xrightarrow{\text{id}} & \mathbb{Z}[\varphi] & \xrightarrow{\text{id}} & \mathbb{Z}[\varphi] & \xrightarrow{\text{id}} & \mathbb{Z}[\varphi] & \longrightarrow & \cdots
\end{array}$$

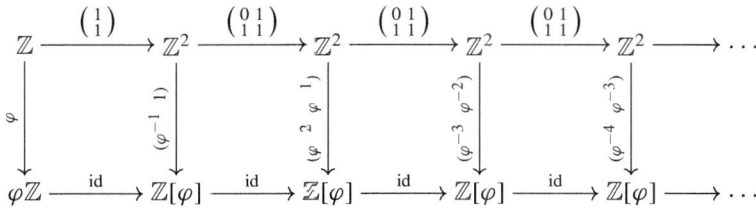

Here, if b is an $\ell \times k$ integer matrix, by $\mathbb{Z}^k \xrightarrow{b} \mathbb{Z}^\ell$ we mean the map given by matrix multiplication by b.

For $k \ge 1$, the $(k+1)$th vertical arrow is the map

$$(m, n) \mapsto \varphi^{-k}(m + n\varphi)$$

and is an isomorphism onto $\varphi^{-k}(\mathbb{Z} + \varphi\mathbb{Z})$, since φ is irrational. Commutativity of each square comes from the property (A.4) of the golden ratio. The maps in the first row are isomorphisms as well, since the matrix (6.10) is invertible in $M_2(\mathbb{Z})$. Finally, for each $k \ge 0$ one has

$$\varphi^{-k}(\mathbb{Z} + \varphi\mathbb{Z}) = \mathbb{Z}[\varphi].$$

Indeed, on one side $\mathbb{Z}[\varphi] = \mathbb{Z} + \varphi\mathbb{Z}$, since $\mathbb{Z} + \varphi\mathbb{Z}$ is a sub-ring of $\mathbb{Z}[\varphi]$ and contains its generators 1 and φ. On the other side, multiplication by φ is an isomorphism in $\mathbb{Z} + \varphi\mathbb{Z}$ represented, in the basis $\{1, \varphi\}$, by the matrix (6.10).

The direct limit of the second row of the diagram gives

$$K_0(A) \cong \mathbb{Z}[\varphi].$$

The positive cone is $\bigcup_{k\geq 1} \varphi^{-k}(\mathbb{N} + \varphi\mathbb{N})$ and it is obviously contained in the set

$$\{x \in \mathbb{Z}[\varphi] : x \geq 0\} = \mathbb{Z}[\varphi] \cap [0, \infty[.$$

We now show that this inclusion is an equality. Let $x = m_0 + n_0\varphi \geq 0$, $m_0, n_0 \in \mathbb{Z}$. We want to show that, for k big enough, $\varphi^k x =: m_k + n_k\varphi$ with $m_k, n_k \geq 0$.

For $x = 0$ the statement is obviously true. Let $x > 0$. From (A.10) we see that

$$m_k = F_{k-1}m_0 + F_k n_0, \qquad n_k = F_k m_0 + F_{k+1}n_0.$$

Since $F_k/F_{k-1} \to \varphi$, there exists k_0 such that $|F_k/F_{k-1} - \varphi| < x/n_0$ for all $k \geq k_0$. Therefore, for all $k \geq k_0$, the ratios

$$m_k/F_{k-1} = x + (F_k/F_{k-1} - \varphi)\, n_0 \qquad n_k/F_k = x + (F_{k+1}/F_k - \varphi)n_0 \tag{6.17}$$

are positive, hence $m_k, n_k > 0$.

Finally, the scale is

$$\bigcup_{k\geq 1} \left\{\varphi^{-k}(m + \varphi n) : 0 \leq m \leq F_{k-1}, 0 \leq n \leq F_k\right\}$$

Elements in the kth set are bounded by $\varphi^{-k}(F_{k-1} + \varphi F_k) = 1$. Thus, the scale is contained in the set $\mathbb{Z}[\varphi] \cap [0, 1]$. We must show that this inclusion is an equality.

Let $x = m_0 + n_0\varphi \in [0, 1]$ and define $\varphi^k x =: m_k + n_k\varphi$ as before. We already proved that for k big enough $m_k, n_k \geq 0$. By a similar argument, if $x < 1$ there exists k_1 such that $|F_k/F_{k-1} - \varphi| < (1 - x)/n_0$ for all $k \geq k_1$. From (6.17) we deduce that $m_k/F_{k-1} \leq 1$ and $n_k/F_k \leq 1$ for all $k \geq k_1$. Finally, if $x = 1$ then $0 \leq m_0 = 1$ and $0 = n_0 \leq 1$. ■

Notice that the dimension groups (6.16) and (6.15) look very similar, but one is finitely generated (as an abelian group) and the other is not.

The idea of computing the triple (6.12) for an AF algebra is due to George Elliott [Ell76], and is part of his famous classification program for C*-algebras.

6.4 AF Equivalence Relations

In the previous chapter we learned how to compute the K-theory of AF algebras, and in fact we computed explicitly the K-theory of the two algebras that are of interest for us: the CAR C*-algebra and the Fibonacci C*-algebra. In this section we turn full circle and show that these two algebras are groupoid C*-algebras of the equivalence relations in (6.1). In fact, we will illustrate the general construction that realizes an arbitrary AF algebra as a groupoid C*-algebra.

The first step is to associate an equivalence relation to an arbitrary Bratteli diagram (cf. Definition 6.14).

Definition 6.21 An *infinite path* in a Bratteli diagram $\mathfrak{B} = (V, E, s, t)$ is a denumerable sequence $(e_0, e_1, \dots, e_n, \dots)$ of edges satisfying

$$e_i \in E_i \quad \text{and} \quad t(e_i) = s(e_{i+1}),$$

for all $i \geq 0$. The set of all infinite paths in $\mathfrak{B}$ is denoted by $X_{\mathfrak{B}}$.

Note that we only consider infinite paths starting from the single vertex of V_0.

Example 6.22 Consider the Bratteli diagram of the CAR C*-algebra:

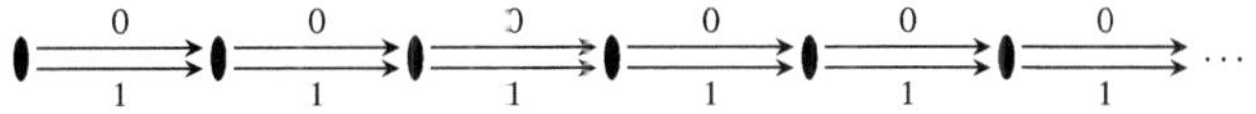

Here every set V_i is a singleton, and every set E_i has two elements, that we call 0 and 1 for obvious reasons. At every vertex we can choose arbitrarily one of the two edges to move forward. Thus, $X_{\mathfrak{B}} = 2^{\mathbb{N}}$ is the set of all binary sequences.

Example 6.23 Consider the Bratteli diagram of the Fibonacci C*-algebra:

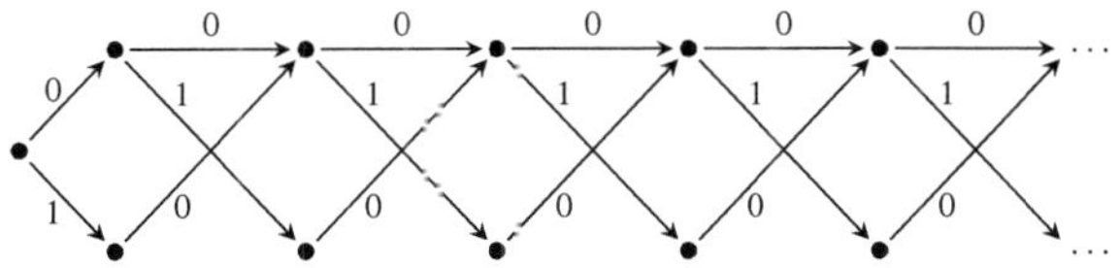

We label by 0 the edges with target on the first row and by 1 the edges with target on the second row. In this way, infinite paths are in bijection with binary sequences in which each 1 is followed by a 0. That is, $X_{\mathfrak{B}} = \mathbb{S}$ is the set of index sequences of Penrose tilings.

We shall denote by $R_{\mathfrak{B}}$ the relation of tail equivalence on $X_{\mathfrak{B}}$: given two infinite paths $x = (x_i)_{i \geq 0}$ and $y = (y_i)_{i \geq 0}$ in $X_{\mathfrak{B}}$, one has

$$(x, y) \in R_{\mathfrak{B}} \iff \exists\, n \geq 0 : x_i = y_i \ \forall\, i \geq n.$$

For future use, let us also define finite paths.

Definition 6.24 A *path of length* $n \geq 1$ in a Bratteli diagram $\mathfrak{B}$ is an n-tuple $(e_0, e_1, \dots, e_{n-1})$ of edges such that $t(e_i) = s(e_{i+1})$ for all $0 \leq i < n-1$. The vertex $v := s(e_0)$ is called the *source* of the path, $w := t(e_{n-1})$ is its *target*, and the path is said to go *from* v *to* w. (An example is in Fig. 6.3.)

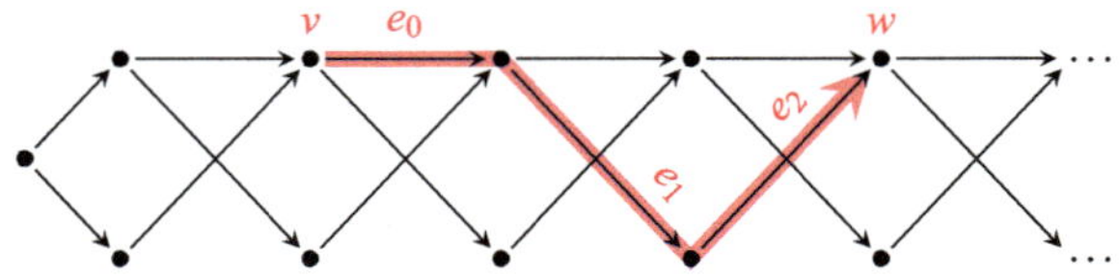

Fig. 6.3 A finite path from v to w

Remark 6.25

(i) The condition that a Bratteli diagram has no sources guarantees that, for every $w \in V \setminus V_0$, there exists a path from $v_0 \in V_0$ to w (by a simple induction, since for $n \geq 1$, every vertex in V_n is the target of an edge starting from V_{n-1}).

(ii) The condition that it has no sinks guarantees that every finite path can be extended to an infinite path.

After recalling the definition of full groupoid C*-algebra $C^*(R)$ of an étale equivalence relation R (Sect. 6.4.1), we will show that there are natural topologies on $X_{\mathfrak{B}}$ (Sect. 6.4.2) and on $R_{\mathfrak{B}}$ (Sect. 6.4.3) such that the relation is étale and $C^*(R_{\mathfrak{B}})$ is isomorphic to the AF algebra with Bratteli diagram $\mathfrak{B}$ (Sect. 6.4.4). In other words, we will construct the two missing maps in the picture

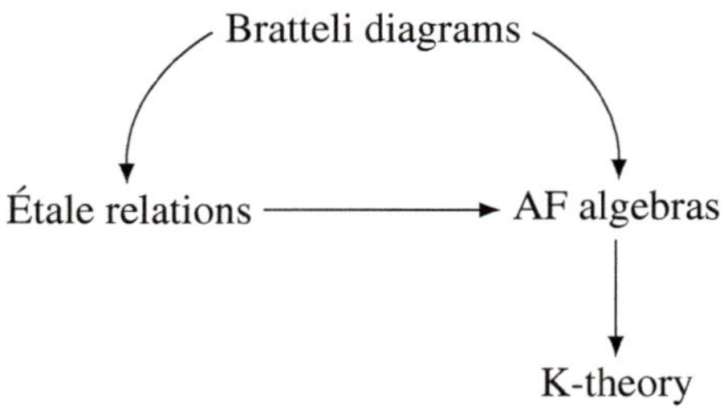

and show that this diagram is commutative.

A reference for this section is the beautiful monograph [Put18]. An equivalence relation coming from a Bratteli diagram is called an *AF relation*.

6.4.1 Étale Relations and Their Convolution Algebra

An *étale* relation R on a topological space X is an equivalence relation $R \subseteq X \times X$ equipped with a topology (usually not the subspace topology from $X \times X$ but a finer one) such that the projection onto the first component

$$\mathrm{pr}_1 : R \to X, \qquad (x, y) \mapsto x,$$

is a local homeomorphism. A subset S of an equivalence relation R is called a *bisection* if there exists an open neighborhood U of S such that $\mathrm{pr}_1 : U \to X$ is a homeomorphism onto its image, which is required to be open. Thus R is étale if and only if it is covered by open bisections.

Lemma 6.26 *If X is a Hausdorff space and $R \subset X \times X$ an étale equivalence relation, the space $C_c(R)$ of compactly-supported complex-valued continuous functions on R is an associative *-algebra, with product and involution given by:*

$$(f \star g)(x, y) := \sum_{z \in [x]} f(x, z) g(z, y) \tag{6.18a}$$

$$f^*(x, y) := \overline{f(y, x)} \tag{6.18b}$$

for all $(x, y) \in R$ and all $f, g \in C_c(R)$.

Proof We wish to show that only finitely many elements in the sum (6.18a) are non-zero, so that the sum is well-defined. It is then enough to show that the set

$$(\mathrm{pr}_1)^{-1}(x) \cap \mathrm{supp}(f) \tag{6.19}$$

is finite. But local homeomorphisms have discrete fibers and, since points in X are closed, $(\mathrm{pr}_1)^{-1}(x)$ is closed and discrete. Thus (6.19) is the intersection of a closed discrete subset and a closed compact subset of R, which means that it is a finite set.

The rest of the proof is elementary. In particular, associativity follows from transitivity of R. (For a generalization to relations that are not transitive, one can see [DLL22].) ∎

To pass to C*-algebras we need some extra conditions. We can think of an equivalence relation as a groupoid $R \rightrightarrows X$, with source and target maps given by the projections onto the first and second component, respectively, and composition and inversion given by:

$$(x, y) \cdot (y, z) := (x, z), \qquad (x, y)^{-1} := (y, x).$$

Equip the set of composable pairs with the topology inherited from the product topology on $R \times R$. Then, $R \rightrightarrows X$ is a topological groupoid if the following maps are continuous: (1) the composition, as a map from composable pairs to R; (2) the inversion, as a map $R \to R$; (3) the map $\mathrm{Id} : X \to R$, $x \mapsto \mathrm{Id}_x := (x, x)$; (4) the projections $\mathrm{pr}_1, \mathrm{pr}_2 : R \to X$ onto both the first and second component.

Observe that if X is Hausdorff and $\mathrm{pr}_1 : R \to X$ is a local homeomorphism, then R is Hausdorff as well. If X is also compact, by the closed map lemma $\mathrm{Id} : X \to R$ is a homeomorphism onto its image, given by the diagonal $\Delta := \{(x, x) : x \in X\}$, which is then compact w.r.t. the topology inherited from R. Thus, when X is compact Hausdorff and R étale, it is not the topology of X that determines the topology of R (R is usually not a topological subspace of $X \times X$), but rather the converse: the topology on R determines the one on X via the homeomorphism with the diagonal.

Lemma 6.27 *If $R \rightrightarrows X$ is a topological groupoid, X is compact Hausdorff, and R is étale and sequential,*[1] *then the algebra $(C_c(R), \star)$ is unital with unit element η given by the function*

$$\eta(x, y) = \begin{cases} 1 & \text{if } x = y, \\ 0 & \text{otherwise} \end{cases}$$

Proof By hypothesis Δ is compact. Since R is Hausdorff, Δ is closed. We now prove that Δ is open, so that its characteristic function η is continuous and compactly supported. The fact that it is a neutral element for $\star$ is a simple computation.

Let $f := \mathrm{Id} \circ \mathrm{pr}_1 : R \to \Delta$, observe that f is the identity on the diagonal, and that it is a local homeomorphism (w.r.t. the topology on Δ inherited from R).

By contradiction, assume that $R \smallsetminus \Delta$ is not closed. Since R is a sequential space, it follows that $R \smallsetminus \Delta$ is not sequentially closed. Thus, there exists a sequence of points (x_n, y_n) in $R \smallsetminus \Delta$ with limit $(x, x) \in \Delta$. Since f is continuous, $f(x_n, y_n) = (x_n, x_n) \to f(x, x) = (x, x)$ as well. Fix an open neighborhood U of (x, x) in R. For n big enough, both (x_n, y_n) and (x_n, x_n) belong to U. Since $(x_n, y_n) \notin \Delta$, one has $(x_n, y_n) \neq (x_n, x_n)$. But $f(x_n, y_n) = (x_n, x_n) = f(x_n, x_n)$. Hence $f|_U$ is not injective. Since U is an arbitrary open neighborhood of (x, x) in R, this proves that f is not locally injective at (x, x) and then not a local homeomorphism. ■

There are several ways to complete $(C_c(R), \star)$ to a C*-algebra. We only mention the basic idea behind the *full groupoid C*-algebra* $C^*(R)$. We need some technical assumptions: the spaces must be Hausdorff, locally compact and second countable. If R is étale, the condition of second countability guarantees that equivalence classes are countable sets (finite or infinite). Then, for $f \in C_c(R)$ we define $\|f\|_{\max}$ as the sup of the operator norm $\|\pi(f)\|$ over all bounded *-representations π of $(C_c(R), \star)$. This set of representations is non-empty, because of the zero representation. The technical conditions guarantee that $\|f\|_{\max}$ is finite for all f (see Proposition 9.2.1 of [SSW20]). It is a C*-seminorm, since each $\|\pi(\,.\,)\|$ is a C*-seminorm. Finally, one proves that $\|\,.\,\|_{\max}$ is a norm, which is done by constructing an injective representation of $C_c(R)$ called the *left regular representation*, see [SSW20, Sect. 9.3]. The completion of $C_c(R)$ in the norm $\|\,.\,\|_{\max}$ is the full groupoid C*-algebra $C^*(R)$.

[1] In the examples we are interested in, R will be second-countable, which implies sequential. Cf. Proposition 6.39(i).

A crucial property of this C*-norm, whose proof can be found in [SSW20, Proposition 9.2.1], is the following.

Lemma 6.28 *Suppose $R \rightrightarrows X$ is a locally compact, Hausdorff, second countable, étale groupoid. Then, for every $f \in C_c(R)$ supported on a bisection, one has*

$$\|f\|_{\max} \leq \|f\|_\infty$$

(where the one on the right hand side is the sup norm).

We shall not define the *reduced groupoid C*-algebra $C^*_r(R)$* (see e.g. [SSW20]), but simply mention that for relations coming from Bratteli diagrams the full and reduced groupoid C*-algebras coincide.

6.4.2 Cantor Spaces of Infinite Paths

Now, we go back to Bratteli diagrams. As before, let $X_{\mathfrak{B}}$ be the set of infinite paths in a Bratteli diagram $\mathfrak{B} = (V, E, s, t)$. A metric d on $X_{\mathfrak{B}}$, generalizing (6.2), is given by the formula

$$d(x, y) := \inf\left\{2^{-n} : n \in \mathbb{N} \text{ and } x_i = y_i \ \forall\, 0 \leq i < n\right\}. \tag{6.20}$$

Note that $d(x, y) = 0$ if and only if $x = y$, and $d(x, y) = 1$ if and only if $x_0 \neq y_0$.

Lemma 6.29 *The distance* (6.20) *is an* ultrametric, *i.e. it satisfies the* strong triangle inequality

$$d(x, z) \leq \max\left\{d(x, y), d(y, z)\right\}.$$

Proof We can assume that x, y, z are all different (otherwise the inequality is trivially satisfied). Let $d(x, y) = 2^{-m}$ and $d(y, z) = 2^{-n}$, with $m, n \in \mathbb{N}$. Then $x_i = y_i = z_i$ for all $i < \min\{m, n\}$ and $d(x, z) \leq \max\{2^{-m}, 2^{-n}\}$. ■

We equip $X_{\mathfrak{B}}$ with the topology induced by this metric, a basis of which is given by open balls. Since d has image in the set $\{2^{-n} : n \in \mathbb{N}\} \cup \{0\}$, it is enough to consider balls whose radius is in this set.

For $n \in \mathbb{N}$ and $x \in X_{\mathfrak{B}}$, we define

$$\mathbb{B}(x, n) := \left\{y \in X_{\mathfrak{B}} : y_i = x_i \ \forall\, 0 \leq i < n\right\}. \tag{6.21}$$

For $n = 0$, the side condition is void and $\mathbb{B}(x, 0) = X_{\mathfrak{B}}$ for every path x. Clearly $\mathbb{B}(x, n)$ depends only on the first n edges of x.

It is straightforward to verify that, for all $n \in \mathbb{N}$ and $x, y \in X_{\mathfrak{B}}$, one has

$$d(x, y) < 2^{-n+1} \iff y \in \mathbb{B}(x, n).$$

The cylinder sets (6.21) are then our open balls. In (6.21), any $2^{-n} < r \leq 2^{-n+1}$ is the radius, and x is the center. In fact, a center, since in an ultrametric space every point inside a ball is its center:

$$z \in \mathbb{B}(x, n) \iff \mathbb{B}(x, n) = \mathbb{B}(z, n). \tag{6.22}$$

It is a general property of ultrametric spaces that open balls are also closed, and that two balls have either empty intersection or one is contained in the other. We verify explicitly these statements in the present case.

We shall use the term *clopen* as an abbreviation of "closed and open".

Lemma 6.30 *Let $m, n \in \mathbb{N}$ and $x, y \in X_{\mathfrak{B}}$. Then:*

(i) The balls $\mathbb{B}(x, m)$ and $\mathbb{B}(y, n)$ are either disjoint or concentric. In the latter case, if $m \geq n$ then $\mathbb{B}(x, m) \subseteq \mathbb{B}(y, n)$, and the inclusion is an equality if $m = n$.
(ii) The collection

$$\mathfrak{P}_n := \{\mathbb{B}(p, n) : p \in X_{\mathfrak{B}}\}$$

is a finite partition of $X_{\mathfrak{B}}$.
(iii) $\mathbb{B}(x, n)$ is clopen.

Proof (i) Assume that $m \geq n$. If there exists $z \in \mathbb{B}(x, m) \cap \mathbb{B}(y, n)$, it follows from (6.22) that the two balls are concentric with center z: $\mathbb{B}(x, m) = \mathbb{B}(z, m)$ and $\mathbb{B}(y, n) = \mathbb{B}(z, n)$. The bigger one is then contained in the smaller one: $\mathbb{B}(z, m) \subseteq \mathbb{B}(z, n)$. If $m = n$, this inclusion is clearly an equality.
(ii) From point (i), we see that balls with fixed radius are pairwise disjoint. If $z \in X_{\mathfrak{B}}$, clearly $z \in \mathbb{B}(z, n)$, proving that the sets in $\mathfrak{P}_n$ cover $X_{\mathfrak{B}}$. The cardinality of $\mathfrak{P}_n$ is at most equal to the cardinality of $\bigcup_{i=0}^{n-1} E_i$. Thus, $\mathfrak{P}_n$ is a finite set.
(iii) From point (ii), $X_{\mathfrak{B}} \smallsetminus \mathbb{B}(x, n)$ is the union of all the balls in $\mathfrak{P}_n$ different from $\mathbb{B}(x, n)$, hence it is open. ∎

Like all metric spaces, $X_{\mathfrak{B}}$ is Hausdorff. Moreover, $\bigcup_{n \geq 0} \mathfrak{P}_n$ is a basis for its topology and is a countable union of finite sets, hence $X_{\mathfrak{B}}$ is second-countable.

Proposition 6.31 *The space $X_{\mathfrak{B}}$ is compact.*

Proof For a metric space, compactness and sequential compactness are equivalent. We shall prove the latter. Let $x^{(n)} = (x_0^{(n)}, x_1^{(n)}, \ldots)$, $n \in \mathbb{N}$, be a sequence of infinite paths in $\mathfrak{B}$ (a sequence of sequences).

Recall that each set of arrows E_i is finite. Since $(x_0^{(n)})_{n \in \mathbb{N}}$ is a sequence in a finite set, at least one element of the set appears infinitely many times in the sequence. There is then an infinite subset $S_0 \subseteq \mathbb{N}$ such that $(x_0^{(n)})_{n \in S_0}$ is a constant sequence. Next, since $(x_1^{(n)})_{n \in S_0}$ is a sequence in a finite set, there is an infinite subset $S_1 \subseteq S_0$ such that $(x_1^{(n)})_{n \in S_1}$ is a constant sequence. And so on.

By induction we construct a descending chain of infinite sets

$$\mathbb{N} \supseteq S_0 \supseteq S_1 \supseteq \ldots \supseteq S_k \supseteq \ldots$$

such that $(x_i^{(n)})_{n \in S_k}$ is a constant sequence for all $0 \leq i \leq k$.

There exists a strictly increasing sequence $(n_k)_{k \geq 0}$ with $n_k \in S_k$ for all $k \geq 0$. Indeed, given $n_k \in S_k$, if S_{k+1} has no element greater than n_k, it would mean that $S_{k+1} \subseteq \{0, \ldots, n_k\}$ is finite, which is a contradiction.

Given any strictly increasing sequence $(n_k)_{k \geq 0}$ as above, we now prove that the subsequence $(x^{(n_k)})_{k \geq 0}$ of our original sequence in $X_{\mathfrak{B}}$ is convergent to the infinite path $y = (y_0, y_1, \ldots)$ defined by $y_i := x_i^{(n_i)}$ for all $i \geq 0$.

For starters, observe that y is indeed a path in $\mathfrak{B}$: for every $i \geq 0$ and every $n \in S_{i+1}$, from $y_i = x_i^{(n)}$ and $y_{i+1} = x_{i+1}^{(n)}$ we deduce that the target of y_i is equal to the source of y_{i+1} (since $x^{(n)}$ is a path).

Since $x_i^{(n_k)} = y_i$ for all $0 \leq i \leq k$, we deduce that

$$d(x^{(n_k)}, y) \leq 2^{-k-1},$$

and the right hand side goes to 0 for $k \to \infty$. ■

Now that we have a topology on $X_{\mathfrak{B}}$, we can ask whether the relation $R_{\mathfrak{B}}$ is minimal (Definition 6.3). The next proposition gives a necessary and sufficient condition.

Let W be a subset of vertices in the Bratteli diagram $\mathfrak{B}$. We say that W is *hereditary* if, for all edges e with $s(e) \in W$, one has $t(e) \in W$ as well (in other words, you can never escape from W moving in the direction of the edges). We say that W is *saturated* if, for all $v \in V$,

$$t\big(s^{-1}(v)\big) \in W \implies v \in W.$$

Thus, W is saturated if every vertex emitting only edges that ends in W is itself contained in W. The trivial subsets $\varnothing$ and V are both hereditary and saturated.

Remark 6.32 By induction, one proves that:

(i) if W is hereditary, $w \in W$ and there is a finite path (of any length) from w to a vertex v, then $v \in W$ as well;
(ii) if W is saturated and, for some $n \geq 1$, one has $V_n \subseteq W$, then $V_0 \subseteq W$;
(iii) if W is hereditary and saturated and, for some $n \geq 0$, $V_n \subseteq W$, then $W = V$.

If the only sets of vertices in $\mathfrak{B}$ that are both hereditary and saturated are the trivial ones, i.e. $\varnothing$ and V, then $\mathfrak{B}$ is called *simple*.

The diagrams in Examples 6.22 and 6.23 are simple. A non-simple Bratteli diagram is in Fig. 6.4.

Lemma 6.33 *Let $W \subset V$ be a hereditary and saturated subset and $v \in V_n \setminus W$, $n \geq 0$. Then, there exists $x \in X_{\mathfrak{B}}$ such that $s(x_n) = v$ and $t(x_i) \notin W$ for all $i \geq n$.*

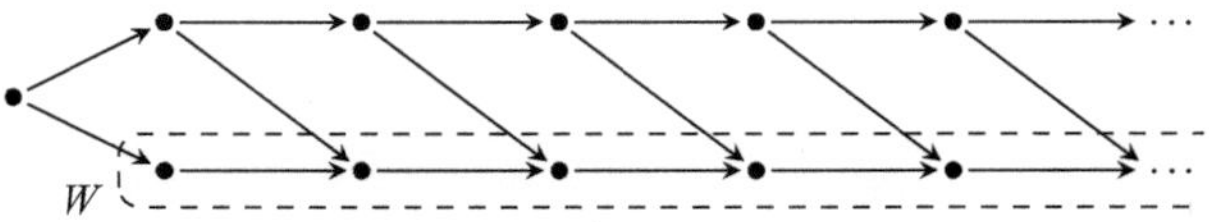

Fig. 6.4 A non-simple Bratteli diagram. The framed set of vertices is hereditary and saturated

Proof We will prove by induction that, for every $m > n$, there is a path $(x_n, \dots, x_{m-1})$ such that $s(x_n) = v$ and $t(x_i) \notin W$ for all $i \geq n$. For $m = n + 1$ the statement is obvious: by contradiction, if every edge from v lands in W, then $v \in W$. Assume that we found a path $(x_n, \dots, x_{m-1})$ as above, for some $m > n$. If every edge from $t(x_{m-1})$ would land in W, we would have $t(x_{m-1}) \in W$, contradicting the inductive hypothesis. Thus, there exists $x_m \in E_m$ such that $s(x_m) = t(x_{m-1})$ and $t(x_m) \notin W$, and $(x_n, \dots, x_m)$ is the desired path of length $m - n + 1$. ■

Proposition 6.34 *The equivalence relation $R_{\mathfrak{B}}$ is minimal if and only if $\mathfrak{B}$ is simple.*

Proof $R_{\mathfrak{B}}$ is minimal if and only if, for all $x, y \in X_{\mathfrak{B}}$ and $n \geq 1$, the intersection:

$$[x] \cap \mathbb{B}(y, n) \tag{6.23}$$

is non-empty. If $t(y_{n-1}) = s(x_n)$, an element in the intersection is

$$(y_0, \dots, y_{n-1}, x_n, x_{n+1}, \dots).$$

The non-trivial case is when $t(y_{n-1}) \neq s(x_n)$, which will be our assumption from now on. In this case, an element z in the intersection (6.23) has the form:

$$z = (y_0, \dots, y_{n-1}, z_n, \dots, z_{m-1}, x_m, x_{m+1}, \dots),$$

for some $m > n$ and for some path $(z_n, \dots, z_{m-1})$ from $t(y_{n-1}) \in V_n$ to $s(x_m) \in V_m$. Of course, such a path may not exist. If there exists a path from a vertex w to a vertex v, we say that v is a *descendant* of w. The intersection (6.23) is empty if and only if no vertex $s(x_m)$ is a descendant of $w = t(y_{n-1})$.

("$\Rightarrow$") If $\mathfrak{B}$ is not simple, pick any non-trivial hereditary and saturated set $W \subset V$, any $w \in W$, call n the integer such that $w \in V_n$ (observe that $n \geq 1$, otherwise it would be $W = V$), and pick any $v \in V_n \smallsetminus W$ (it exists by Remark 6.32). Let $x \in X_{\mathfrak{B}}$ be a path as in Lemma 6.33 (we say that x has its tail outside W), and $y \in X_{\mathfrak{B}}$ any path with $t(y_{n-1}) = w$. Any $z \in \mathbb{B}(y, n)$ has $s(z_n) \in W$, hence $t(z_i) \in W$ for all $i \geq n$ (since W is hereditary). Such a z can never be tail equivalent to x. Thus, we found x, y, n such that the intersection (6.23) is empty, which means that $R_{\mathfrak{B}}$ is not minimal.

An illustration is in Fig. 6.5.

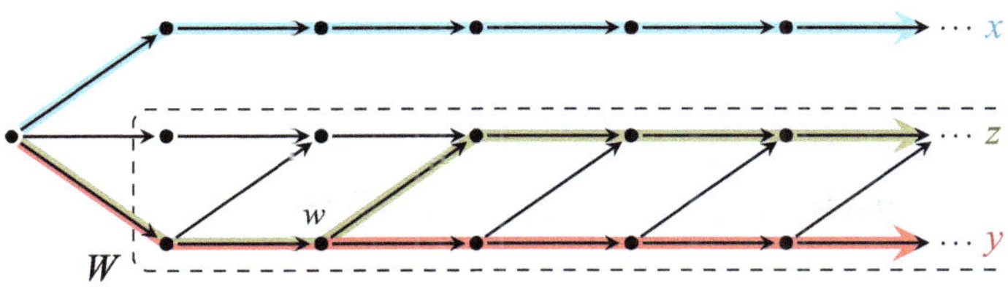

Fig. 6.5 Proof of Proposition 6.34

("$\Leftarrow$") Assume that $R_{\mathfrak{B}}$ is not minimal. Then, there exists $w \in V$ and $x \in X_{\mathfrak{B}}$ such that no vertex $s(x_m)$ is a descendant of w. Let W be the set of vertices v satisfying the following condition:

$$\exists\, i \geq 1 \text{ such that every descendant of } v \text{ in } V_i \text{ is also a descendant of } w.$$

Since $s(x_m) \notin W$ (for all m), one has $W \neq V$. Since $w \in W$, one has $W \neq \varnothing$.

By construction, W is hereditary: let $e \in E_k$ be an edge with source $v \in W$ and target $u \in V_{k+1}$. For some i, every descendant of v in V_i is a descendant of w. If $i > k+1$, every descendant of u in V_i is also a descendant of v, and then of w, proving that $u \in W$. If $i = k+1$, then u is a descendant of w, and we reach the same conclusion.

Finally, W is saturated: let $u \in V_k$ be a vertex emitting only edges ending in W. Let $i := k+2$. Every descendant of u in V_i is a descendant of a vertex in W, hence a descendant of w. This proves that $u \in W$. ∎

We now explain the reason for the title of this subsection. Let us recall a few well-known facts from General Topology. We refer to [Wil04] for the details.

A topological space is called *zero-dimensional* if it has a basis of clopen sets. Every zero-dimensional Hausdorff space is totally disconnected.

A topological space is *perfect* if it has no isolated points (i.e. no open singletons).

Definition 6.35 A *Cantor space* is a non-empty topological space that is: compact, metrizable, totally disconnected and perfect.

The next theorem is a rephrasing of [Wil04, Corollary 30.4].

Theorem 6.36 *[Brouwer] There is only one Cantor space, up to homeomorphism: the Cantor set.*

We already know that $\mathbb{S}$ and $2^{\mathbb{N}}$ are homeomorphic to the Cantor set. The next proposition generalizes this to simple Bratteli diagrams, with the only exception of the trivial cases where there is only one tail equivalence class.

Proposition 6.37 *If $\mathfrak{B}$ is a simple Bratteli diagram, then either $X_{\mathfrak{B}}$ is a Cantor space or $X_{\mathfrak{B}}/R_{\mathfrak{B}}$ is a one-point space (i.e. $R_{\mathfrak{B}} = X_{\mathfrak{B}} \times X_{\mathfrak{B}}$).*

Proof $X_{\mathfrak{B}}$ is compact (Proposition 6.31) and zero-dimensional (Lemma 6.30(iii)), hence totally disconnected. If it has no isolated points, it is a Cantor space.

Assume, then, that $X_{\mathfrak{B}}$ has (at least) one isolated point y, i.e. that $\{y\}$ is open. Every open neighborhood of y contains a set $\mathbb{B}(y, n)$ for n big enough (remember that every point inside a ball is its center), hence $\{y\} = \mathbb{B}(y, n)$ for some value of n. Since $R_{\mathfrak{B}}$ is minimal (Proposition 6.34), for every $x \in X_{\mathfrak{B}}$ the intersection $[x] \cap \mathbb{B}(y, n) = [x] \cap \{y\}$ is non-empty, that means $(x, y) \in R_{\mathfrak{B}}$. Since every point is in relation with a fixed y, there is only one equivalence class. ■

6.4.3 From Bratteli Diagrams to AF Relations

We adopt the notations of the previous subsections. In particular, $\mathfrak{B} = (V, E, s, t)$ is a Bratteli diagram (we do not assume that it is simple). For $n \geq 1$ and $v \in V_n$, we denote by Γ_v the set of all paths x in $\mathfrak{B}$ of length at least n (possibly infinite) that start from $v_0 \in V_0$ and such that x_{n-1} has target v. We denote by Γ_{v_0} the set of all paths starting from v_0 (finite or infinite).

For all $n \geq 0$, $v \in V_n$ and $p, q \in \Gamma_v$ we define

$$U(p, q, n) := \big\{(x, y) \in \mathbb{B}(p, n) \times \mathbb{B}(q, n) : x_i = y_i \ \forall\, i \geq n\big\}. \tag{6.24}$$

Note that $U(p, q, 0)$ is the diagonal of $X_{\mathfrak{B}} \times X_{\mathfrak{B}}$ (for all p, q).

Lemma 6.38 *Two sets $U(p, q, n)$ and $U(p', q', n')$ are either disjoint or contained one into the other; if $n = n'$, they are either disjoint or equal.*

Proof The proof is similar to that of Lemma 6.30 and we omit it. ■

Proposition 6.39

(i) The collection of sets

$$\Big\{U(p, q, n) : n \geq 0,\ \ v \in V_n,\ \ p, q \in \Gamma_v\Big\}$$

is a countable basis for a topology on $R_{\mathfrak{B}}$.

W.r.t. this topology:

(ii) $R_{\mathfrak{B}}$ is Hausdorff.
(iii) For all $(x, y) \in R_{\mathfrak{B}}$, the collection of sets (6.24) *containing (x, y) is a neighborhood basis at (x, y).*
(iv) The subspace topology on $U(p, q, n)$ induced from $R_{\mathfrak{B}}$ coincides with the subspace topology inherited from the product topology on $X_{\mathfrak{B}} \times X_{\mathfrak{B}}$.
(v) The sets (6.24) *are compact. Thus, $R_{\mathfrak{B}}$ is locally compact.*

Proof (i) Firstly, we want to show that the sets (6.24) form a cover:

$$R_{\mathfrak{B}} = \bigcup_{n\geq 0} \bigcup_{v\in V_n} \bigcup_{p,q\in\Gamma_v} U(p,q,n).$$

That is, we want to prove that x and y are tail equivalent if and only if $(x, y) \in U(p, q, n)$ for some p, q, n. Both implications are trivial: if $(x, y) \in U(p, q, n)$, then x and y are tail equivalent by definition of the set; if $x_i = y_i$ for all $i \geq n_0$ (x and y are tail equivalent), then $(x, y) \in U(x, y, n_0)$.

The second condition for a basis of a topology follows from Lemma 6.38, since the intersection of any two sets $U(p, q, n)$ and $U(p', q', n')$ is either empty or equal to one of the two sets.

The basis at point (i) is countable. Indeed, for each $n \geq 0$, the set V_n is finite. Hence, for each $v \in V_n$, there are only finitely many paths of length n in Γ_v (by convention, the only path of length zero is the empty path).
(ii) Let $(x, y), (x', y') \in R_{\mathfrak{B}}$. Then, there exists $n_0 \in \mathbb{N}$ such that, for all $n \geq n_0$,

$$(x, y) \in U(x, y, n) \subset \mathbb{B}(x, n) \times \mathbb{B}(y, n),$$
$$(x', y') \in U(x', y', n) \subset \mathbb{B}(x', n) \times \mathbb{B}(y', n).$$

If, for some $n \geq n_0$, $\mathbb{B}(x, n) \cap \mathbb{B}(x', n) = \varnothing$ or $\mathbb{B}(y, n) \cap \mathbb{B}(y', n) = \varnothing$, then $U(x, y, n)$ and $U(x', y', n)$ are disjoint open sets separating the points (x, y) and (x', y'). If, for all $n \geq n_0$, $\mathbb{B}(x, n) \cap \mathbb{B}(x', n) \neq \varnothing$ and $\mathbb{B}(y, n) \cap \mathbb{B}(y', n) \neq \varnothing$, then it must be $x = x'$ and $y = y'$.
(iii) Given an arbitrary union S of sets (6.24) and a point $(x, y) \in S$, any set $U(p, q, n)$ containing (x, y) (and there exists at least one) is entirely contained in S (because of Lemma 6.38).
(iv) From Lemma 6.38, a basis for the topology on $U(p, q, n)$ induced from $R_{\mathfrak{B}}$ is given by those $U(x, y, k)$ that are subsets of $U(p, q, n)$. This happens if $k \geq n$, $x_i = p_i$ and $y_i = q_i$ for all $0 \leq i < n$, $x_i = y_i$ for all $n \leq i < k$. In such a case, from

$$U(x, y, k) = U(p, q, n) \cap \big(\mathbb{B}(x, k) \times \mathbb{B}(y, k)\big)$$

we see that the subspace topologies on $U(p, q, n)$ induced from $R_{\mathfrak{B}}$ and $X_{\mathfrak{B}} \times X_{\mathfrak{B}}$ coincide.
(v) Because of point (iv), since $X_{\mathfrak{B}} \times X_{\mathfrak{B}}$ is compact, it is enough to show that $U(p, q, n)$ is closed in $X_{\mathfrak{B}} \times X_{\mathfrak{B}}$. But $U(p, q, n)$ is the intersection of the closed set $\mathbb{B}(p, n) \times \mathbb{B}(q, n)$ with the set

$$\big\{(x, y) \in X_{\mathfrak{B}} \times X_{\mathfrak{B}} : x_i = y_i \ \forall\, i \geq n\big\}. \tag{6.25}$$

The complement of the set (6.25) is a union $\bigcup_{i\geq n} Y_i$, with

$$Y_i := \big\{(x, y) \in X_{\mathfrak{B}} \times X_{\mathfrak{B}} : x_i \neq y_i\big\}.$$

It remains to show that each set Y_i is open, so that (6.25) is closed, and $U(p, q, n)$ is closed. But

$$Y_i := \bigcup_{x, y \in X_{\mathfrak{B}} : x_i \neq y_i} \mathbb{B}(x, i+1) \times \mathbb{B}(y, i+1),$$

hence the thesis. ■

From now on, we shall think of $R_{\mathfrak{B}}$ as a topological space with the topology generated by the sets (6.24).

Theorem 6.40 *With the above topologies, $R_{\mathfrak{B}} \rightrightarrows X_{\mathfrak{B}}$ is a (Hausdorff, locally compact, second-countable) étale groupoid. Each set* (6.24) *is a bisection.*

Proof First, we must show that all structure maps are continuous. For every topological space X, the trivial relation $R := X \times X$ with product topology gives a topological groupoid $R \rightrightarrows X$. In particular, $X_{\mathfrak{B}} \times X_{\mathfrak{B}} \rightrightarrows X_{\mathfrak{B}}$ is a topological groupoid.

Even if $R_{\mathfrak{B}}$ is not a topological subspace of $X_{\mathfrak{B}} \times X_{\mathfrak{B}}$, its topology agrees with the subspace topology on arbitrarily large subsets, and this is enough to prove that the structure maps of the former groupoid are continuous. More precisely, let $(x, y) \in R_{\mathfrak{B}}$ and choose n big enough so that $(x, y) \in U(x, y, n)$. The inversion map

$$X_{\mathfrak{B}} \times X_{\mathfrak{B}} \to X_{\mathfrak{B}} \times X_{\mathfrak{B}}, \qquad (x, y) \mapsto (y, x),$$

gives by restriction a continuous map $U(x, y, n) \mapsto U(y, x, n)$. The inversion in $R_{\mathfrak{B}}$ is then locally continuous, which means that it is continuous. In the same way one proves the continuity of the source and target maps.

The map $\mathrm{Id} : X_{\mathfrak{B}} \to X_{\mathfrak{B}} \times X_{\mathfrak{B}}$ is a homeomorphism onto the diagonal $U(p, q, 0)$. On the diagonal, the subspace topologies induced from $X_{\mathfrak{B}} \times X_{\mathfrak{B}}$ and $R_{\mathfrak{B}}$ coincide, hence $\mathrm{Id} : X_{\mathfrak{B}} \to R_{\mathfrak{B}}$ is continuous.

Assume that x, y, z are all tail equivalent, and choose n big enough so that both $(x, y) \in U(x, y, n)$ and $(y, z) \in U(y, z, n)$. The multiplication of composable pairs in $X_{\mathfrak{B}} \times X_{\mathfrak{B}}$ restricts to a continuous map

$$U(x, y, n) \times U(y, z, n) \to U(x, z, n),$$

proving that the multiplication of composable pairs in $R_{\mathfrak{B}}$ is locally continuous, hence continuous.

Finally, for all p, q, n, the map

$$\mathrm{pr}_1 : U(p, q, n) \to \mathbb{B}(p, n) \tag{6.26}$$

is bijective. Indeed, if $x = (p_0, \dots, p_{n-1}, x_n, x_{n+1}, \dots) \in \mathbb{B}(p, n)$ is any point in the codomain, there is a unique $y \in X_{\mathfrak{B}}$ such that $(x, y) \in U(p, q, n)$, and is given by $y = (q_0, \dots, q_{n-1}, x_n, x_{n+1}, \dots)$. Since (6.26) is bijective, continuous and with compact domain, it is a homeomorphism. Since the restriction of pr_1 to any basic open set of $R_{\mathfrak{B}}$ is a homeomorphism onto its image, the groupoid is étale. ■

Example 6.41 Let $\mathfrak{B}$ be one of the two Bratteli diagrams in Examples 6.22–6.23.

Let $x = (0, 0, 0, \dots)$ be the null sequence and $y = (1, 0, 1, 0, 0, 0, \dots)$ the sequence with 1's in first and third position and zeros everywhere else. We now show that $U(x, y, 2)$ is not open in $X_{\mathfrak{B}} \times X_{\mathfrak{B}}$ in the product topology, proving that with the topology generated by the sets (6.24) $R_{\mathfrak{B}}$ is not a topological subspace of $X_{\mathfrak{B}} \times X_{\mathfrak{B}}$.

For $n \geq 0$, let $x^{(n)} \in X_{\mathfrak{B}}$ be the sequence with 1 in position n and zeros everywhere else. Then, $(x^{(n)}, y) \notin U(x, y, 2)$, either because $x^{(n)} \notin \mathbb{B}(x, 2)$ for $n \leq 2$, or because $x^{(n)}$ and y have different nth component for $n \geq 3$. On the other hand, $d(x^{(n)}, x) = 2^{-n} \to 0$ for $n \to \infty$, thus $(x^{(n)}, y) \to (x, y) \in U(x, y, 2)$ in the product topology. This proves that the complement of $U(x, y, 2)$ in $X_{\mathfrak{B}} \times X_{\mathfrak{B}}$ is not sequentially closed, and so $U(x, y, 2)$ is not open.

6.4.4 From AF Relations Back to AF Algebras

We now move on to investigate the groupoid C*-algebra of the groupoid $R_{\mathfrak{B}} \rightrightarrows X_{\mathfrak{B}}$ defined in the previous subsection. We shall keep using the same notations.

For $n \geq 0$, $v \in V_n$ and $p, q \in \Gamma_v$ of length exactly n, we denote by $E^n_{p,q}$ the characteristic function of the set $U(p, q, n)$:

$$E^n_{p,q}(x, y) := \begin{cases} 1 & \text{if } (x, y) \in U(p, q, n) \\ 0 & \text{otherwise.} \end{cases}$$

By convention, the only path of length zero is the empty path, and $E^0_{p,q}$ is the characteristic function of the diagonal $U(p, q, 0)$.

Forcing the labels to be paths of lengths exactly n gives us bijections between triples p, q, v, sets $U(p, q, n)$, and functions $E^n_{p,q}$.

Since $U(p, q, n)$ is clopen and compact, $E^n_{p,q}$ is a continuous compactly-supported function on $R_{\mathfrak{B}}$. Clearly, the involution (6.18b) satisfies

$$(E^n_{p,q})^* = E^n_{q,p}.$$

The convolution product (6.18a) of two such characteristic functions behaves like the product of matrix units.

Lemma 6.42 *For all $n \geq 0$, $v, v' \in V_n$, $p, q \in \Gamma_v$ and $p', q' \in \Gamma_{n,v'}$ of length n, one has*

$$E^n_{p,q} \star E^n_{p',q'} = \delta_{v,v'} \delta_{q,p'} E^n_{p,q'}.$$

Proof The product is supported on the set $U(p, q, n) \cap U(p', q', n')$, which is empty if $v \neq v'$. If $v = v'$, for all $(x, y) \in R_{\mathfrak{B}}$ one has

$$(E^n_{p,q} \star E^n_{p',q'})(x, y) = \sum_{z\in[x]} E^n_{p,q}(x, z)E^n_{p',q'}(z, y).$$

To get a non-zero summand we need both $(x, z) \in U(p, q, n)$ and $(z, y) \in U(p', q', n)$, which implies $p' = (z_0, \ldots, z_{n-1}) = q$, and $x_i = z_i = y_i$ for all $i \geq n$.

If this is the case, since z is uniquely determined by p', q, x, the sum is 1. Therefore, when $v = v'$ and $q = p'$, then $(E^n_{p,q} \star E^n_{p',q'})(x, y)$ is 1 if $(x, y) \in U(p, q', x)$ and is 0 otherwise, proving that $E^n_{p,q} \star E^n_{p',q'} = E^n_{p,q'}$. ■

For each $n \geq 0$, call

$$A_n := \text{span}\{E^n_{p,q} : v \in V_n,\ p, q \in \Gamma_v \text{ of length } n\}.$$

For each $n \geq 0$ and $v \in V_n$, let $k(v)$ be the number of paths of length n in Γ_v. Then, from the previous lemma we see that there is an obvious isomorphism of finite-dimensional unital *-algebras

$$A_n \cong \bigoplus_{v\in V_n} M_{k(v)}(\mathbb{C}),$$

sending each $E^n_{p,q}$ to a matrix unit.

In the last proposition, we prove that $C^*(R_{\mathfrak{B}})$ is the AF C*-algebra of the Bratteli diagram $\mathfrak{B}$.

Lemma 6.43

(i) $\bigcup_{n\geq 0} A_n$ *is the set of functions in* $C_c(R_{\mathfrak{B}})$ *with finite range.*
(ii) *Every* $f \in C_c(R_{\mathfrak{B}})$ *supported in a set* $U(p, q, n)$ *is a uniform limit of functions in* $\bigcup_{n\geq 0} A_n$ *supported in the same set* $U(p, q, n)$.
(iii) $\bigcup_{n\geq 0} A_n$ *is dense in* $C^*(R_{\mathfrak{B}})$.

Proof (i) Each $f \in \bigcup_{n\geq 0} A_n$ is a finite linear combination of characteristic functions, hence it may assume only finitely many values.

Conversely, assume that $f \in C_c(R_{\mathfrak{B}})$ has finite range and let $\{a_1, \ldots, a_k\}$ be the set of non-zero values of f. For each $1 \leq i \leq k$, $S_i := f^{-1}(a_i)$ is a clopen set contained in the support of f, hence compact. Clearly

$$f = \sum_{i=1}^{k} a_i \chi_i$$

where χ_i is the characteristic function of S_i. Since S_i is open, it is a union of basic open sets (6.24); since it is compact, it can be written as a finite union of basic open sets (6.24). Finally, since two sets (6.24) are either disjoint or one contained in the other, we can write S_i as a finite *disjoint* union of sets (6.24), and χ_i is a finite sum of characteristic functions $E^n_{p,q}$. Thus $\chi_i \in \bigcup_{n\geq 0} A_n$ and $f \in \bigcup_{n\geq 0} A_n$ as well.

(ii) This follows from the fact that, on a completely regular 0-dimensional space, every bounded continuous functions is a uniform limit of continuous functions with finite range, see e.g. [GJ17, Chap. 16], in particular 16.29(a) and Problem 16A.2 (in the book, all the functions are real-valued, but the result immediately extends to complex-valued functions by considering separately their real and imaginary part). If f has support contained in $U(p, q, n)$, clearly $f = f \cdot E^n_{p,q}$ (where here the product is the pointwise product of functions). But $f = \lim_{k\to\infty} f_k$ where each f_k has finite range. Thus

$$f = f \cdot E^n_{p,q} = \lim_{k\to\infty} (f_k E^n_{p,q})$$

is also the uniform limit of continuous functions $f_k E^n_{p,q}$ with finite range and support contained in $U(p, q, n)$. From point (i) it follows that $f_k E^n_{p,q} \in \bigcup_{n\geq 0} A_n$.
(iii) Let $f \in C_c(R_{\mathfrak{B}})$. We can cover the support of f with basic open sets (6.24). Since the support of f is compact, we can extract a finite subcover $\mathcal{U} = \{U_\alpha\}$. Let $\{\rho_\alpha\}$ a partition of unity subordinated to the cover $\mathcal{U}$, then

$$f = \sum_\alpha \rho_\alpha f$$

is a finite sum of functions $\rho_\alpha f$ each supported on a set (6.24). From point (ii), every such a function is a uniform limit of functions in $\bigcup_{n\geq 0} A_n$ supported in the same basic open set. But the sets in (6.24) are bisections, and Lemma 6.28 implies that, for a sequence of functions supported on a bisection, uniform convergence implies norm-convergence in $C^*(R_{\mathfrak{B}})$. Thus, $\bigcup_{n\geq 0} A_n$ is norm-dense in $C_c(R_{\mathfrak{B}})$, that is norm-dense in $C^*(R_{\mathfrak{B}})$. ■

Proposition 6.44

(i) $A_0 = \mathbb{C}I$.
(ii) For every $n \geq 0$, A_n *is a unital* C^**-subalgebra of* A_{n+1}.
(iii) The AF algebra $C^*(R_{\mathfrak{B}}) = \overline{\bigcup_{n\geq 0} A_n}$ *has Bratteli diagram* $\mathfrak{B}$.

Proof (i) is trivial. A_0 is spanned by the single element $E^0_{\varnothing,\varnothing}$, the characteristic function of the diagonal, which is the neutral element of the convolution product.
(ii) follows from the observation that, for all $n \geq 0$, $v \in V_n$ and $p, q \in \Gamma_v$ of length n, one has

$$U(p, q, n) = \bigcup_{e\in E_n : s(e)=v} U(pe, qe, n + 1)$$

where $pe = (p_0, \ldots, p_{n-1}, e)$ denotes the concatenation of two paths. Using this, a straightforward computation gives

$$E^n_{p,q} = \sum_{e\in E_n : s(e)=v} E^{n+1}_{pe,qe}.$$

From this we see that $A_n \subseteq A_{n+1}$, and in fact that we have an inclusion of a factor $M_{k(v)}(\mathbb{C})$ of A_n into a factor $M_{k(v')}(\mathbb{C})$ of A_{n+1} whenever there is an edge e connecting v with v'. This proves (iii).

All algebras A_n share the same unit, given by:

$$\sum_{v \in V_n, p \in \Gamma_v} E^n_{p,p} = E^0_{\varnothing,\varnothing}.$$

The inclusions $A_n \subseteq A_{n+1}$ are then unital. ■

We close with a comment. By definition, $\| \, . \, \|_{\max}$ is the biggest C*-norm on the convolution algebra $(C_c(R_{\mathfrak{B}}), \star)$. The inequality in Lemma 6.28 then holds for any C*-norm on $(C_c(R_{\mathfrak{B}}), \star)$. One can repeat the proof of Lemma 6.43 to conclude that, if A is any C*-algebra completion of $(C_c(R_{\mathfrak{B}}), \star)$, then $\bigcup_{n \geq 0} A_n$ is dense in A. But on each finite-dimensional C*-algebra A_n there is a unique C*-norm, thus on the dense subalgebra $\bigcup_{n \geq 0} A_n$ the norm of A coincides with $\| \, . \, \|_{\max}$, proving that $A = C^*(R_{\mathfrak{B}})$. Therefore, the convolution algebra has a unique C*-completion, given by the full groupoid C*-algebra.

Appendix A
Some Useful Formulas

A.1 The Golden Ratio

Throughout the book, we denote by ζ the primitive 5th root of unity:

$$\zeta := e^{2\pi i/5},$$

that we identify with the unit vector

$$\left(\cos\tfrac{2\pi}{5}, \sin\tfrac{2\pi}{5}\right)$$

under the usual identification of $\mathbb{C}$ and $\mathbb{R}^2$. Consider now the path:

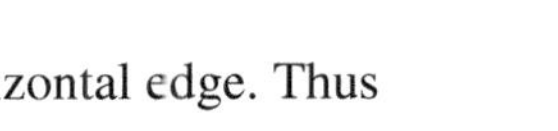

(A.1)

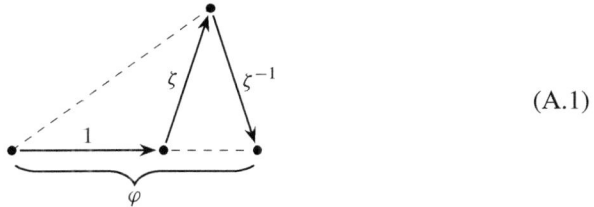

Call φ the length of the horizontal edge. Thus

$$\varphi = 1 + \zeta + \zeta^{-1}. \tag{A.2}$$

Throughout the book we use repeatedly the fact that, for all $n \notin 5\mathbb{Z}$:

$$\sum\nolimits_{j=0}^{4} \zeta^{nj} = 0, \tag{A.3}$$

which follows from the formula for the sum of a geometric series,

F. D'Andrea, *A Guide to Penrose Tilings*,
https://doi.org/10.1007/978-3-031-28428-1

$$\sum_{j=0}^{4} \zeta^{nj} = (1-\zeta^{5n})/(1-\zeta^n),$$

and the fact that $\zeta^5 = 1$. In particular, (A.3) for $n = 1$ tells us that

$$\varphi^2 - \varphi - 1 = 1 + \zeta + \zeta^2 + \zeta^{-2} + \zeta^{-1} = 1 + \zeta + \zeta^2 + \zeta^3 + \zeta^4 = 0.$$

Thus, φ solves the quadratic equation:

$$\varphi^2 = \varphi + 1, \tag{A.4}$$

whose positive solution is the *golden ratio*:

$$\varphi = \frac{1+\sqrt{5}}{2}.$$

Observe that the triangle (A.1) has one edge of unit length, since ζ is a unit vector, and the remaining one of length $|1+\zeta|$, which is still equal to φ. The internal angle are easily computed: the one on the bottom-right corner is $2\pi/5$ and since it is an isosceles triangle, the remaining angles must be $2\pi/5$ and $\pi/5$.

A triangle which is similar to (A.1) will be called a *golden triangle*.

Next, consider next the isosceles triangle with unit length legs:

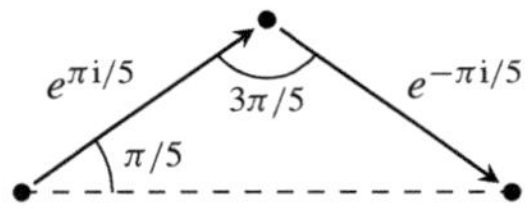

(A.5)

The square of the base is given by $(e^{\pi i/5} + e^{-\pi i/5})^2 = \zeta + \zeta^{-1} + 2 = \varphi + 1 = \varphi^2$, proving that the base has length φ.

A triangle which is similar to (A.5) will be called a *golden gnomon*.

Note that in a isosceles triangle the base is twice the length of the legs times the cosine of the angle at the base. From this observation, applied to the triangles (A.1) and (A.5), we derive the following useful trigonometric formula:

$$\cos\frac{\pi}{5} = \frac{1}{2}\varphi, \qquad \cos\frac{2\pi}{5} = \frac{1}{2}\varphi^{-1}. \tag{A.6}$$

Finally, we can compute the area of the above triangles, which is given by the length of a leg times half-base times the sine of the angle at the base. If we denote by L the triangle in (A.1) and S the triangle in (A.5), after some manipulation with (A.4), we get:

$$\text{Area}(S) = \frac{1}{4}\sqrt{\varphi+2}, \qquad \text{Area}(L) = \varphi \cdot \text{Area}(S). \tag{A.7}$$

In the picture (A.1) we see the decomposition of the golden triangle with unit base into a golden gnomon with unit legs and a smaller golden triangle, with unit legs and base $\varphi - 1 = \varphi^{-1}$ (thus, L rescaled by φ^{-1}). Similarly, a golden gnomon decomposes into smaller golden triangle and golden gnomon (rescaled by φ^{-1}):

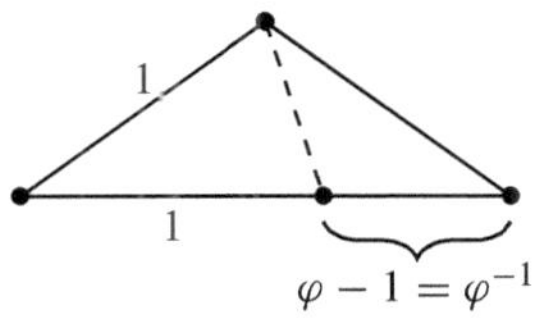

We write these decompositions symbolically as

$$L = S + \varphi^{-1} L, \qquad S = \varphi^{-1} S + \varphi^{-1} L. \tag{A.8}$$

Inside the triangle S, and then inside L, we can inscribe a circle:

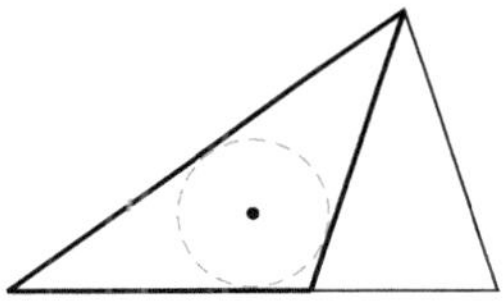

of radius

$$\rho = \frac{1}{2\sqrt{\varphi + 2}}.$$

(The radius of the inner circle of a triangle is given by the area of the triangle over the semi-perimeter.) Since $\varphi < 2$, we have $\rho > 1/4$.

A.2 Fibonacci Numbers

From (A.4) it follows that:

$$\varphi^{n+2} = \varphi^{n+1} + \varphi^n \qquad \forall\, n \geq 0. \tag{A.9}$$

If, for $n \geq 1$, we write

$$\varphi^n = F_n \varphi + F_{n-1} \tag{A.10}$$

it follows that (A.9) is satisfied if the coefficients in the above equation satisfy the recurrence relation:

$$F_{n+2} = F_{n+1} + F_n \ \forall\, n \geq 0, \quad F_0 = 0, \quad F_1 = 1.$$

That is, F_n is the nth *Fibonacci number*. In deriving (A.10) we only used the fact that φ is a solution of the equation $x^2 = x + 1$. A similar identity, then, holds for the other solution $-\varphi^{-1}$ of that equation:

$$(-\varphi^{-1})^n = F_n(-\varphi^{-1}) + F_{n-1}. \tag{A.11}$$

Subtracting (A.11) from (A.10) one finds *Binet's formula*:

$$F_n = \frac{\varphi^n - (-\varphi^{-1})^n}{\varphi + \varphi^{-1}} = \frac{\varphi^n - (-\varphi^{-1})^n}{\sqrt{5}}, \tag{A.12}$$

which trivially holds for $n = 0$ as well. Observe that, since $\varphi > 1$, in Binet's formula $\varphi^n \to \infty$ and $(-\varphi^{-1})^n \to 0$ for $n \to \infty$. Thus

$$\lim_{n\to\infty} F_{n+1}/F_n = \varphi. \tag{A.13}$$

By induction on $n \geq 1$ we can prove the matrix identity

$$\begin{pmatrix} F_{n+1} & F_n \\ F_n & F_{n-1} \end{pmatrix} = \begin{pmatrix} 1 & 1 \\ 1 & 0 \end{pmatrix}^n. \tag{A.14}$$

This is obviously true for $n = 1$, and the inductive step is simply the recursive relation rewritten in matrix form:

$$\begin{pmatrix} F_{n+1} & F_n \\ F_n & F_{n-1} \end{pmatrix} = \begin{pmatrix} 1 & 1 \\ 1 & 0 \end{pmatrix} \begin{pmatrix} F_n & F_{n-1} \\ F_{n-1} & F_{n-2} \end{pmatrix}.$$

Taking the determinant of both sides of (A.14) one arrives at *Cassini's identity*:

$$F_{n+1}F_{n-1} - (F_n)^2 = (-1)^n. \tag{A.15}$$

Finally, using (A.12), one can prove that

$$\frac{F_{2k}}{F_{2k-1}} < \varphi < \frac{F_{2k+1}}{F_{2k}} \tag{A.16}$$

for all $k \geq 1$.

A.3 Cut and Practice with Robinson Triangles

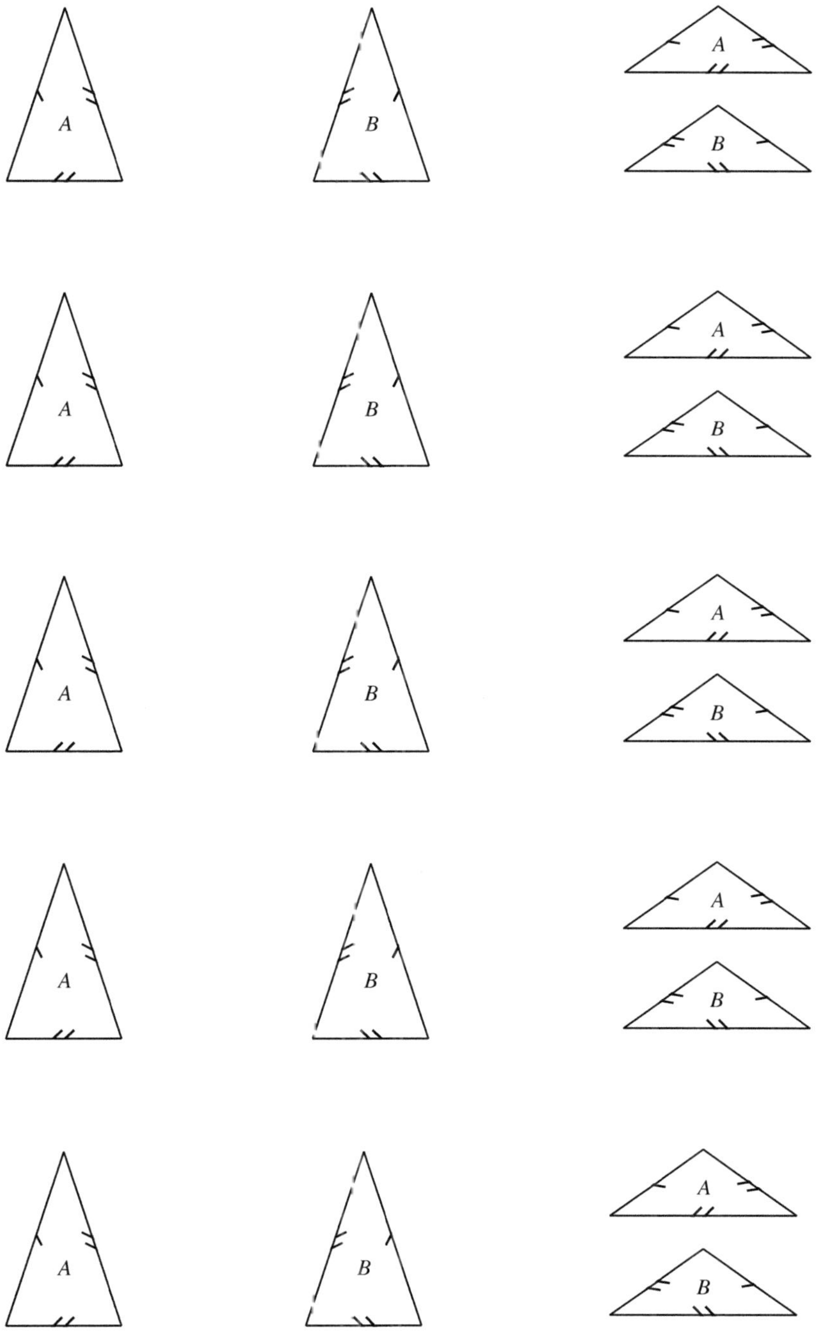

References

[AYP13] H. Au-Yang, J.H.H. Perk, Quasicrystals—the impact of N.G. de Bruijn. Indag. Math. **24**(4), 996–1017 (2013)

[Bab01] R. Babilon, 3-colourability of Penrose kite-and-dart tilings. Discret. Math. **235**(1–3), 137–143 (2001)

[BBI01] D. Burago, Y. Burago, S. Ivanov, *A Course in Metric Geometry*. Graduate Studies in Mathematics, vol. 33 (AMS, 2001)

[Bee93] G. Beer, *Topologies on Closed and Closed Convex Sets* (Kluwer, 1993)

[Bel86] J. Bellissard, K-theory of C*-algebras, in *solid state physics, in Statistical Mechanics and Field Theory: Mathematical Aspects*. ed. by T.C. Dorlas, N.M. Hugenholtz, M. Winnik, L. Notes, in Physics, vol. 257, (Springer, Berlin, 1986), pp.99–156

[Ber66] R. Berger, *The Undecidability of the Domino Problem*. Number 66 in Mem. Amer. Math. Soc. AMS (1966)

[BFG07] M. Baake, D. Frettlöh, U. Grimm, A radial analogue of Poisson's summation formula with applications to powder diffraction and pinwheel patterns. J. Geom. Phys. **57**(5), 1331–1343 (2007)

[Bla86] B. Blackadar, *K-Theory for Operator Algebras*, vol. 5 (MSRI Publications, Springer, 1986)

[Bur94] A. Burns, Fractal tilings. Math. Gaz. **78**(482), 193–196 (1994)

[Con94] A. Connes, *Noncommutative Geometry* (Academic Press, 1994)

[Dav96] K.R. Davidson, *C*-Algebras by Example*, The Fields Institute Monographs, vol. 6 (AMS, 1996)

[dB81] N.G. de Bruijn, Algebraic theory of Penrose's non-periodic tilings of the plane. I, II. Indag. Math. **43**(1), 39–66 (1981)

[dBE51] N.G. de Bruijn, P. Erdős, A colour problem for infinite graphs and a problem in the theory of relations. Indag. Math. **13**, 371–373 (1951)

[DGS82] L. Danzer, B. Grünbaum, G.C. Shephard, Can all tiles of a tiling have five-fold symmetry? Am. Math. Mon. **89**(8), 568–585 (1982)

[DLL22] F. D'Andrea, G. Landi, F. Lizzi, Tolerance relations and quantization. Lett. Math. Phys. **112**(4), 1–28 (2022)

[Ell76] G.A. Elliott, On the classification of inductive limits of sequences of semisimple finite-dimensional algebras. J. Algebr. **38**(1), 29–44 (1976)

[Fer18] A. Fernholm, Crystals of golden proportions (2018). https://www.nobelprize.org/uploads/2018/06/popular-chemistryprize2011.pdf

F. D'Andrea, *A Guide to Penrose Tilings*,
https://doi.org/10.1007/978-3-031-28428-1

[FH13] D. Frettlöh, E. Harriss, Parallelogram tilings, worms, and finite orientations. Discret. Comput. Geom. **49**(3), 531–539 (2013)

[FHG] D. Frettlöh, E. Harriss, F. Gähler, Tilings encyclopedia. https://tilings.math.uni-bielefeld.de/

[Gar77] M. Gardner, Mathematical games. Sci. Am. **236**(1), 110–121 (1977)

[Gar97] M. Gardner, *Penrose Tiles to Trapdoor Ciphers: And the Return of Dr (Matrix* (Cambridge University Press, Cambridge, 1997)

[GBVF13] J.M. Gracia-Bondía, J.C. Várilly, H. Figueroa, *Elements of Noncommutative Geometry* (Springer, Berlin, 2013)

[GJ17] L. Gillman, M. Jerison, *Rings of Continuous Functions* (Courier Dover Publications, 2017)

[GS87] B. Grünbaum, G.C. Shephard, *Tilings and Patterns* (Courier Dover Publications, 1987)

[HSW10] M. Holz, K. Steffens, E. Weitz, *Introduction to Cardinal Arithmetic* (Birkhäuser, 2010)

[JR15] E. Jeandel, M. Rao, An aperiodic set of 11 Wang tiles. Adv. Comb. (2015). https://doi.org/10.19086/aic.18614

[Kla23] E. Klarreich, Hobbyist finds math's elusive 'Einstein' tile. Quantamagazine (2023). https://www.quantamagazine.org/hobbyist-finds-maths-elusive-einstein-tile-20230404/

[Knu69] D.E. Knuth, *The Art of Computer Programming/ 1* (SIAM, Fundamental Algorithms, 1969)

[KP00] J. Kellendonk, I.F. Putnam, Tilings, C*-algebras and K-theory, in *Directions in Mathematical Quasicrystals*, ed. by M. Baake and R.V. Moody. CRM Monograph Series, vol. 13 (AMS, 2000)

[Lan03] G. Landi, *An Introduction to Noncommutative Spaces and Their Geometries* (Springer, Berlin, 2003)

[Lay82] S.R. Lay, *Convex Sets and Their Applications* (Wiley, 1982)

[Min98] L. Minnick, *Generalized Forcing in Aperiodic Tilings* (B.sc. thesis, Williams College, Williamstown, Massachusetts, USA, 1998)

[Mul20] D. Muller, The infinite pattern that never repeats (2020). https://youtu.be/48sCx-wBs34

[OAC] Martin Gardner Papers (SC0647). Dept. of Special Collections and University Archives, Stanford University Libraries, Stanford, Calif. https://oac.cdlib.org/findaid/ark:/13030/kt6s20356s/

[Pen74] R. Penrose, The role of aesthetics in pure and applied mathematical research. Bull. Inst. Math. Appl. **10**, 266–271 (1974)

[Pen78] R. Penrose, Pentaplexity. Eureka **39**, 16–22 (1978)

[Pen89] R. Penrose, Tilings and quasi-crystals; a non-local growth problem? in *Introduction to the Mathematics of Quasicrystals (Aperiodicity and Order)*, Chap. 2 (Academic Press, 1989), pp. 53–79

[Put18] I.F. Putnam, *Cantor Minimal Systems*. University Lecture Series, vol. 70 (AMS, 2018)

[Rao17] M. Rao, Exhaustive search of convex pentagons which tile the plane (2017). arXiv:1708.00274

[Ren06] J. Renault, *A Groupoid Approach to C*-Algebras*. Lecture Notes in Mathematics, vol 793 (Springer, Berlin, 2006)

[Rie82] M.A. Rieffel, Applications of strong Morita equivalence to transformation group C*-algebras, in Proceedings of Symposia in Pure Mathematics, vol. 38 (1982)

[Rob75] R.M. Robinson, Comments on the Penrose tiles. Mimeographed Notes, preprint of the University of California Berkeley (1975)

[Sad08] L.A. Sadun, *Topology of Tiling Spaces*. University Lecture Series, vol. 46 (AMS, 2008)

[Sch78] D. Schattschneider, Tiling the plane with congruent pentagons. Math. Mag. **51**(1), 29–44 (1978)

[Sen95] M. Senechal, *Quasicrystals and Geometry* (Cambridge University Press, Cambridge, 1995)

[SMKGS23] D. Smith, J.S. Myers, C.S. Kaplan, C. Goodman-Strauss, An aperiodic monotile (2023). arXiv:2303.10798

[SSW20] A. Sims, G. Szabó, D. Williams, *Operator Algebras and Dynamics: Groupoids, Crossed Products, and Rokhlin Dimension* (Springer, Berlin, 2020)

[ST11] J.E.S. Socolar, J.M. Taylor, An aperiodic hexagonal tile. J. Comb. Th. A **118**(8), 2207–2231 (2011)

[ST12] J.E.S. Socolar, J.M. Taylor, Forcing nonperiodicity with a single tile. Math Intel. **1**(34), 18–28 (2012)

[SW00] T. Sibley, S. Wagon, Rhombic Penrose tilings can be 3-colored. Am. Math. Mon. **107**(3), 251–253 (2000)

[Wan61] H. Wang, Proving theorems by pattern recognition. II. Bell Syst. Tech. J. **40**(1), 1–41 (1961)

[Wil04] S. Willard, *General Topology* (Dover Publications, 2004)

[Zon20] C. Zong, Can you pave the plane with identical tiles? Not. AMS **67**(5) (2020)